OIL NOTES

OIL
NOTES

RICK BASS

Drawings by Elizabeth Hughes

Houghton Mifflin/Seymour Lawrence

Boston · 1989

For information about permission to reproduce selections
from this book, write to Permissions, Houghton Mifflin
Company, 2 Park Street, Boston, Massachusetts 02108.

Library of Congress Cataloging-in-Publication Data

Bass, Rick, date.
Oil notes.
1. Bass, Rick, date.   2. Petroleum geologists—
United States—Biography.   3. Oil well drilling.
I. Title.
TN869.2.B37A3   1989   622′.3382′0924 [B]   88-32820
ISBN 0-395-48675-0

Printed in the United States of America

Book design by Robert Overholtzer

Q 10 9 8 7 6 5 4 3 2 1

For Elizabeth

In the book of Job, the Lord demands, "Where wast thou when I laid the foundations of the earth? Declare, if thou hath understanding! Who laid the cornerstones thereof, when the morning stars sang together, and all the sons of God shouted for joy?"

"I was *there!*"—surely that is the answer to God's question. For no matter how the universe came into being, most of the atoms in these fleeting assemblies that we think of as our bodies have been in existence since the beginning.

—Peter Matthiessen,
The Snow Leopard

Acknowledgments

Grateful acknowledgment is made to the publications in which excerpts of this book first appeared—*Antaeus* and *Southwest Review*. Editorial advice on this project has come from Leslie Wells, Camille Hykes, and Larry Cooper, to whom I am much indebted. Two geologists have been particularly instrumental and patient in educating me about geology throughout my career—my father, C. R. Bass, and Mr. Ev Ellison—and this debt is also acknowledged.

OIL NOTES

In the fall of 1985, my twenty-seventh year, I read in a work by the poet-novelist Jim Harrison a quote from Kafka about "freeing the frozen sea within us."

I know how to find oil, but I'm a horrible speaker: I couldn't sell men's magazines on a troopship. And I don't know if I can even write well enough to explain why oil is found in some places and not in others. I get frustrated. It seems sometimes that the best way to communicate the presence of oil—or perhaps of anything—is to revert to guttural ughs and growls, and just go out and, by damn, sink a hole in the ground, shove the pipe down there deep enough, until oil begins to flow up out of it, bubbling, with its rich smell of hiddenness and with the energy of discovery. And then to point to it: to say, There it is. Always, I want to do that. I want biceps to sheen; I want tractor-trailers to groan, bringing materials in and taking oil out, and drilling breaks to squeal. (You're drilling through a hard formation, bearing down; then the drill bit pierces a softer formation, one that is more capable of storing oil. The pipe shifts, sinking down into this softer formation, going faster, and it makes a barking, torquing, squealing sound. It sounds exactly like beagles . . .)

I want to stamp on the ground hard enough to make that oil come out. I want to skip legalities, permits, red tape, and other obstacles.

Sometimes I feel almost out of control, and that what is down there is between the oil and me. I want to go immediately and straight to what matters: getting that oil.

My father calls me Animal. I was a fence post in the

third-grade play. I bump into things often, and frequently run over others. But I know where oil is, and I want to try to explain to you what it feels like, how it is, to know this.

I just do not know how to do it—show you—because it is three-dimensional, or even beyond. It is future, undrilled and I am present, knowing. I don't know yet, without drilling, how to bridge that gap.

It is the frozen sea within me.

I know how to find oil.

The great jump o' the mouse. (Morning, wilderness, air; summer was in it.) I was driving back from Elizabeth's, up the Natchez Trace (north out of Rocky Springs), and I saw this little field mouse go scampering across the black Trace in front of me, and he gathered himself for a tremendous leap and made it into the weeds safely.

I am a geologist, and these are going to be notes that I write in little journal books I carry around. Not all of them are going to be about oil and not all of them are going to be about Elizabeth and a whole lot of them are going to lack any structure at all, but if you know a geologist, you know that that is the way he expresses things. Notes: there is no continuity in a geologist's life, not in an active, busy one, anyway. Elizabeth is the woman I've been dating for about four years now, and the Trace, the federal parkway between her house in the country and the town of Jackson, Mississippi, site of my office, is perhaps the prettiest drive

downhill from the Rockies. It was April, the morning (sipping coffee, 50 mph speed limit; no traffic, solitary) I saw the mouse and marveled at his leap. The mouse reminded me of a geologist's life, that busy, busy scampering and then the leap. And my not being able to tell, in the end, if he had to make that leap, or if he was just doing it because it was morning and it was spring. It was a pretty good leap, at any rate.

But in the spring, on the Trace, you can't watch for mice too much; you've got to keep a lookout. And you can't gaze at the fields and barns and palominos knee-deep in grass as you whiz by, because lined all up and down the road like checkpoints are the box turtles, visible as high, shiny glints even a long way off, on a sunny morning, if you are watching. Their shells look like little anticlines, which is where you want to drill, of course, if you are lucky enough to find such a thing. An anticline is a dome-shaped structure beneath the earth; a buried hill, sort of, where a formation rises to its highest elevation. Since oil floats on water, and there is water (called formation water, usually remnant water from ancient seas—yes, it tastes salty; yes, drinking too much of it will make you sick) in most formations, the oil, if there is any, will ride on top of that water and float up into the highest part of that formation. Which would be atop the anticline. The crest, we call it.

I am from Houston, Texas, and work in Jackson. I look for oil all over the country. I found some oil in Michigan once. I will entice you, if I can, by telling you this—that I can find oil almost everywhere, if I try hard enough—and thereby

4

explaining, perhaps, why I and other geologists do what we do. (Look for oil.)

It is one of the few things in the world that is not governed by paradox. It's an art, the crudest of the sciences there is, perhaps. It's even called an "earth science"— mud, dirt, rock, greasy old oil, smelly gas; but except for an occasional tilted water table, it usually—hell, almost always—plays by the rules. None of this *i*-before-*e*-except-after-*c* business, or anything like that. It's consistent, and you can count on it, and in general a sturdy kind of man or woman does it. You live or die by your own lights. The rules always stay the same, though.

Not that photographing wild Cape buffalo is an industry, but let's take it as an example. Suppose you are stalking the wily Cape, going after his picture, tromping around in the brush and stuff with one of those camouflage hats on. Suppose that he's very rare, and you see his tracks, and see glimpses of him, but never get a picture. And the sun gets hot, and in the evenings you sit around the fire and drink native drinks and talk about philosophy and the Baltimore Orioles, about hopes for sons and daughters, about mosquitoes, and then in the mornings before daylight you are up and working at it, getting set up in the brush, being silent, then tracking, heart pounding, etc. Well, sure, when you do get the buffalo's picture you are going to feel good—triumphant, proud, and everything—but still there's going to be a letdown, and after it is over, you're going to realize with much certainty that the best part, the absolute best part, was those native drinks and sunrises. And that's how

it is for almost everything. But not for geology, not for finding oil. No sir. Not a wisp of that old paradox.

It's an adrenaline rush, tracking that oil, curling the contours on the maps you draw up and down and around subtle structures, pausing, thinking there may be an anticline below section 36, but then merely nosing, dipping, and going on over into the next township. SP character, depth of invasion, fault traces, gravity data amplitude anomalies wavelets slickensides, an old farmer who swears his water well burns a flare—all these things and about a thousand wonderful more go through your slow old human brain as your thick old human hand continues to drift, gripping a number HB pencil, across your map. Oh yeah, sure, that part's fun, that's the native drink and sunrise part for sure, too.

But when you move the rig in and rig up and drill down and strike the oil . . . there is no letdown.

So now you can tell your geologist friends that being a geologist is not like taking a picture of a buffalo but that at least there is no paradox. One can live by one's lights.

Geologists are odd, sometimes, in different ways. They may just nod and agree.

Elizabeth is my age, twenty-seven, and has long brown hair. She is an artist, and plays tennis pretty well. She likes chocolate ice cream, and is pretty.

Little scratch notes. I am going to tell you what she is like, and then I am going to tell you what the earth is like, and how very, very hard it is to find oil in sufficient quantities to justify the search, and how you may find it in the most unusual and unexpected circumstances.

The place where we like to eat is Crechale's. Every now and again, because she loves to swim in the winter, right after work we will do this: check into the Best Western on Highway 80, for it has, on Wednesday nights, a wonderfully empty (of people) and large heated swimming pool in a green sweating atrium. We'll swim for an hour, go eat at Crechale's, come back to the room and watch a movie. Wake up early and swim again and then have breakfast at the Green Derby (enchiladas, eggs). Go to work. Face Thursday. She is an artist and likes to travel. One day I will live in Colorado or west Texas or Montana and my life will be different than it is now. I feel this urge, and know it, and yet, I am not miserable at all. Places inland where we drill sometimes—these places used to be deep under the Gulf of Mexico, and now the nearest beach is two hundred miles away. I like that. It implies possibilities, potential.

When you drill for oil you cannot help but look past where you are, and down the map, at Other Successes, or, more important—most important—places where Other Successes could possibly be. And then work to make them so.

She, as well as it (oil), has taught me not to apologize for anything that I am. There is me, and there is her. There is me, and then oil is out there too. Everything exists on its own grounds, in its own boundaries.

My job is this: I am a development geologist. Forgive the ego, but I like it better than the job of exploration geologist (the other kind). I'll contrast and compare the two off and on throughout the book, but right now I will tell you that that is my job—to define boundaries. To have responsibility for the Good, and to be aware that in places there is Not Good. I am given a producing well, a new discovery—a wildcat that has proven itself to be good, to be a "producer" ("making a well," they call it: "Did y'all make a well?")—and I must drill all the other wells in the area, or field, after that. I've got to figure out how much oil is there, where it is, how many wells to drill to get it out, and where and when to drill them. The oil's existence has been proven; now I've got to learn that particular reservoir's nature, its characteristics, and the lightning-bolt atrocities of its own underground existence. Found in some places and not in others. Shy here, coy there, blatant elsewhere, and thousands of barrels of pay just a few hundred feet to the north of the discovery well. Sometimes the discovery well isn't the best well in a field. I must find out these things.

There are ways.

Everyone loves to be a teacher. Of course, it's easy to see why this love has survived over the generations: it favors the species, certainly. The magic in learning is no stronger than the pleasure of teaching something: giving a part of yourself, and having someone take it.

There was a bully when I was small who used to beat me up regularly. If I rode my bike through his neighborhood, he would chase me down on foot, push me off the bike, and let me have it. Once by massive misfortune we were on the same baseball team. And there was some rule whereby everyone had to pitch. This athletic bully's team, our team, was winning about three thousand to nothing until it got to be my inning to pitch.

They started catching up real fast.

The bully, whose name was Jimbo, shouted at me from third base. Several times he started toward the mound, ignoring the other team's runners, who kept circling the bases faster and faster. Jimbo banged his fist in his glove and told me all the things that were going to happen to me.

You could see the desperation on his face when this did nothing to improve my pitching. It was like the worst dreams: I wasn't getting the ball near the plate, even after his threats. I *aimed* at the plate, praying that when the batters walloped the ball someone would make a miraculous catch, but my pitches started going behind their backs, over the catcher's head, everywhere. It was as if I had a serious physical disability. The ball was going about halfway between the plate and the dugout, at a forty-degree angle away from where it was supposed to be.

11

Jimbo stormed over and grabbed the ball and put it in my fist. He showed me a way to grip it that I'd never seen done. Then he took it back from me, and with this tremendously hateful, determined scowl, showed me, without actually releasing the ball, how to throw it. He demonstrated in a slow, barely contained, angry motion, the reaching, pulling, aiming.

"Pretend with your free hand like you're pushing someone out of the way," he said. It was easy to see that that was what he was thinking of. His hand concealed the ball until the last moment; then, shoving (slowly, for my benefit) the imaginary, offensive person aside—there was a sudden opening where that person and his free, gloved hand were—like spit and hate, Jimbo threw the ball through there, through that clawed gap of air, a strike, and made the catcher yell "ouch!"

I tried it and threw strikes. He stood there on the mound and watched to make sure I did it right. I was relieved and amazed. I remember his face: he grinned and hid his own amazement. I don't think he had ever taught anyone anything before. I don't think anyone had ever listened.

I threw more strikes. We won the game. The next day Jimbo beat me up. But I knew there was something else in him no one understood: I think I was the first person ever to see it, and was amazed that even an ogre who drubbed me could feel that sensation. The last I heard of him, he was in jail. He couldn't, or wouldn't, spell or write, and in my most tearful moments, tasting my own blood, I would remind him of this. But he could teach you how to throw a baseball better, I suspect, than it has ever been done. There is a tremendous amount of genius in being able to take someone throwing wildly into the dugout and tell him one sentence and have the ball go precisely over the plate.

If I was angry enough, or desperate enough, could I tell someone how to find oil so well that he or she could go out and do it immediately? Could I do it in one sentence? Not even in my most confident moments do I imagine that I could. I do not know what that sentence is. I think that Jimbo was a genius, and hope he got out of jail. The closest I can come to that sentence, beyond "Listen to the earth," is that you have to get down under and beyond the mere occupational greed and look into the simplicity, the purity, the sacred part of it—the act, not the results, and yourself —and be aware that it is history, buried.

Jimbo's sentences were so much shorter. He would scowl at what I just said and beat me up.

Barrels—that's the standard term in the oil industry for measuring fluid. Barrels of oil, barrels of water . . . one barrel equals forty-two gallons. As far as oil wells go, I generally like to think of anything that makes more than a hundred barrels a day as a good well, and anything less than that, a smaller well. Occasionally, people do produce oil wells from shallow depths, at very low rates—three, four, five barrels a day, if the prices are high enough—but you must always be aware of your operating costs. If you have the well "on pump"—hooked up to a rod pump that draws the oil up the casing by suction—that'll be a cost, for the electricity and maintenance of the pump. And then you've got the costs of transporting the oil. And taxes, and lease burdens. And water: sometimes you'll draw water, with your oil, and have to separate the two. You can't sell water, not the old ocean water you sometimes produce from your formation, water from millions of years ago, salty, the ocean then.

You can't just dump that water on the ground, either, not with the little oil drops and filminess in it, and expect it simply to dry out, and not damage anything or make a mess. So you've got to haul it off by truck or pump it down into a saltwater disposal well—usually an abandoned oil well—and into some formation, deep down there, that does not and never will contain drinking water. You put the water back into the earth.

You put the oil back in too, but that is a different story, and happens when you die, when the hydrogens and carbons that you burn go up into the sky, out into the atmosphere as energy dissipated, and come back down in different forms. For a while, anyway . . .

Nothing can truly disappear. It can only be rearranged, so that it *seems* to have disappeared.

The hydrogen and carbon atoms are not smashed; they are not destroyed. Their form is merely altered.

Corn, oak trees, sea kelp, bryozoans, rotifers, algae, rabbits, two-by-fours . . . The carbon goes around the earth, the hydrogen does too, temporary everywhere it goes. Maybe they're one thing for only a few years, a few hundred million years. Sometimes the two link up, the hydrogens to a carbon, and sit down low in the earth and hide as oil. But again: only for a while.

Two terms: wildcat, rank wildcat.

A wildcat is a well drilled in an area where nothing has been found before. There may be an oil field to the south or north of it, or it may even be surrounded by oil fields, but if it's not adjacent to production, rather is striving to establish production—what will almost certainly have to be a new reservoir, if the drillers find something—this is called a wildcat. Wildcats are risky, a little self-indulgent, to my way of thinking, but that's just the way I've been trained. I'd rather find oil than not. It seems all too irresponsible, bordering on dishonesty: spending someone else's money to drill where you are not absolutely *certain* there is pay. I mean, you can tell your investor there are risks, inform him that it's a wildcat, and as such, very likely not to hit, and he'll nod and say, Yes, yes, I understand, but you both know he's still hoping or else he wouldn't be spending the money. I guess I respect that, and am a little more cautious with the subject of hope than I really should be, to be the best businessman I could be.

There's a lot of capacity for heroism, of course, in drilling wildcats, but only if you are able to forget the fact that you bypassed a rough spot in your conscience, back when you didn't know for sure, but went ahead anyway, hoping.

A rank wildcat means it's superwild. It's usually the place you'd try to drill a well if someone told you he'd give you a million dollars to drill a dry hole on purpose. A rank wildcat doesn't only mean there've been some dry holes drilled around it; it means *nothing's* been drilled for maybe

ten, twenty-five, a hundred miles. If wildcats are self-indulgent, then you can imagine what rank wildcats are.

In addition to being lost money, they are necessary.

They're the only way to increase reserves, to find oil that
was not even suspected. You can decrease consumption,
which *relatively* increases the oil your country might have,
but *actually* to increase what you have, you've got to go out
and drill wildcats and rank wildcats. Which is, of course,
what oil-exporting countries would like to see us stop doing.
And which we *are* drastically reducing.

There's a part of me that never wants to drill a dry hole and
a part of me that never wants to let anyone's hopes down,
and this conflicts with the thoughts I have that there is un-
discovered oil in the Lower Tuscaloosa formation of northern Avoyelles Parish; gas in the depths below Kosciusko,
Mississippi; and oil in Bibb County, Alabama. It is necessary that someone try these and other places, and I must be
careful not to judge the big-company geologists who try
areas such as these. But their personalities reflect—sometimes too often—the fact that they don't especially care
about being right or not, and don't particularly mind being
less than careful with people's hopes. We're leaving the
realm of geology here and going into the abrasions of soci-
ology.

—wind on my porch this morning, and a mockingbird
fighting something—John Prine spinning on my turntable—

At any rate, that's what wildcats and rank wildcats are. Development wells (also called field wells, offsets, outposts, confirmation tests) are where you take what you have found and try to make the most of it before going off to another area. A melting pot nonetheless is always best for any cause or purpose, and I must remember again and again not to criticize the other sort of geologist, even if I do not particularly care to be around him.

I hate coal. I won't spend much time on this because my being a petroleum geologist seems to cast an overwhelming bias on the whole thing. But if I weren't dependent on finding and marketing oil and gas, I'd still despise energy generated from coal. I've seen the way it's mined, out west, the things it does to the wildest and most beautiful country. It cuts down the miners in Appalachia, the residents, even. I've seen the lakes up in Vermont and New York turned to sterile pools, like trays in a lab—dead timbers of trees standing in and all around the shores, not even rotting, just dry and dead, acidified, like a graveyard: acid rain.

Coal is filthy. It comes out of buried swamps, which are oxygen-reducing environments, and these swamps, and their coal, contain much sulfur. When the coal is burned, it releases the sulfur into the atmosphere (the air we breathe), which combines with moisture, H_2O, to form sulfuric acid, H_2SO_4. It falls down on our heads, our properties, lives and earth.

There are methods of safely "scrubbing" the sulfur from the coal as it is burned at power plants. (Smokestacks rising high into the air: launching pads for the poison.) The word is "spew"—trying to get the sulfur emissions up into the winds aloft so that they will leave the plant's vicinity. There (dusting of hands: innocent). But our country doesn't enforce the scrubbing requirements of its coal-fired power plants. So, I hate coal.

Fireplaces—the burning of wood, our forests, things cut and killed before their prime for no reason beyond romance —are 90 percent inefficient in the heat they deliver. It all

19

rises: up the chimney, pure waste of tree life and clean air. Walk through the neighborhood on a cold night. Yes, it smells good. Enjoy it while you can.

Nuclear, solar, hydroelectric, and wind power are the only sources cleaner than gas. Nuclear's the only power more volatile than gas. What I am saying is that gas can get the job done, and is a friend, and safe. Good.

This isn't a commercial. This is how it really is. You haven't seen bad until you've seen coal.

Half moon. Feet on railing, under porch light. Thirty-six degrees last night; the same tonight. There's silence, or almost: the freezer hums, faint, and there's one cricket. An owl, too, far away. It's what I like to call silence, anyway. I just got through running enough to sweat—a mile, maybe two. I have moved into a farmhouse in the woods, twenty-four miles from the city. The nearest town, my mailing address, is called Terry. Also in this area of the state are the towns of Crystal Springs and Utica, and, thirty miles away, Vicksburg. I am going to like this place.

I am happier with Elizabeth, far and away for sure I am, but I am not dependent on her for all my happiness. It is a fine line, and yet a broad one. It is like two huge blocks of earth, next to each other—two huge things separated by a tiny fault—with oil on one side of the fault, in one of the blocks of earth, and nothing on the other side, in the other block. Drill in the right block of earth: it all depends on that. Upthrown or downthrown? There are signs, if you look.

I'm sipping orange juice from one of those little cartons with the punch straw; it's all I have. There's also a little two-pound canned ham in the cupboard, one left over from college (seven? eight? years ago) that Mom sent, picturing it as a Sunday treat for me. I was too much into pizza and ice cream cones and Hot Ziggety Dawgs on those Sunday-away-from-the-cafeteria dinners; the ham has followed me back to Texas, to Arkansas, back to Utah, to Mississippi (East Drive, Sykes Park Road, The Grove Apartments, North Street, Cedar Street, and now the farm). Tomorrow

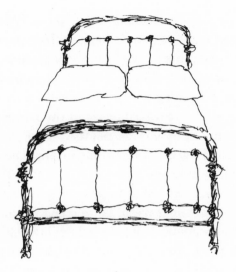

night I may drink a bottle of schnapps and get brave and open it, fry it maybe to make it safer, because for a fact I am hungry. I don't have a refrigerator yet.

The old iron bed is set up. The duck decoy pictures hang on the wall. The bookshelf is set up. It looks very homey. I have yet to spend my first night here. I've got enough stuff moved in so that I can stay tonight if I wish, except for one thing: I have to go out on a well tonight.

Mr. H. has a 10 percent working interest in a thirteen-thousand-foot Lower Tuscaloosa well in Wilkinson County, southeast of Natchez, near Woodville, Mississippi, down in the swamps of the big river, where there are bears. That part of the country is making a lot of new wells in the deeper Lower Tusc after many years of wildly prolific shallow successes in the Wilcox sand. Except that these Wilcox wells (three thousand feet to eight thousand feet deep) make about twenty-five to fifty barrels a day, and some of these Lower Tusc test wells are making over six hundred barrels of oil per day.

Elizabeth is out of town, out of state, traveling; I will go right by her house. The first frost of the season is supposed to be tonight. If she were in, we would fix hot chocolate and maybe play backgammon.

Casey's Place. You think first, when going down to log a well, about what the well will look like, whether it'll be pay or not, but the second thing you think about is where and what you will eat. I am not going to apologize for it. You can't have a passion for one thing and not another. If you like to find oil, you like to eat. I'm not sure I understand it. Only that the feeling for food, oil, and Elizabeth are so often so alike. Casey's Place is a new discovery, about midway between my farmhouse and Elizabeth's, and I am fiercely critical of it, yet sharply interested. The possibility of its being a find is given added excitement by its proximity. It would be like finding an oil well a mile and a half from the office, for this, the nearest café to my house, to be spectacularly good.

The menu has a picture of a pretty decent pizza on it. And there is a beautiful waitress: you cannot look at her too long without aching like you rarely ache. Best to watch her reflection in the window as she moves to a table and asks if anyone needs any more coffee. But enough romantic garbage: forget her jeans and perfect face and dark eyes, black hair and great teeth. Elizabeth is my girl, my honey, and I'm on the road. I like to hold her to me. We have fun together.

The pizza, despite its appearance, is awful.

I drive on.

Elizabeth! Passing her house now. One A.M., two A.M., three: Elizabeth. I think of other things. But she is the only person who holds my interest.

When you drill a well, anything can go wrong with it. The walls are forever trying to cave in. Different formations have various strengths and weaknesses. You need to be aware of the formations.

She laughs when she skates. I think I feel too strongly for her but do not know what to do about it.

The discovery well for South Brush Creek Field—455 barrels per day, in Lamar County, Alabama, up in the far north of the state. The pines a dark green, red clay and steep ridges. My employer and his partners were going to plug it. They said it looked *different*, not like anything they'd ever seen. I said that was because it was oil—before this, only gas had been discovered in the area. They decided to risk a try and not plug it. They ran pipe down into the hole and now it is a field and one of the wells in the field tested 1,067 barrels of oil per day. Twenty-five hundred feet deep (half a mile) and it is making what the deep wells, the fifteen-thousand-footers, make. Ask any geologist about that number: 1,067. Watch his eyes. He might salivate, too.

Now I'm interested in an area a mile away, from there. Actually, I've always been interested in it, but haven't been able to interest my employers. Now they are going to listen. It's been two years. It is like art. It is also like drowning: the feeling of knowing something. You tell them yes, and they say no. Not being able to show it, get it out. The oil, out of the ground, and lots of it.

Elizabeth is different. Elizabeth is oil, is everything. I was interested in her for years. I knew it for a long time. I wanted to do things with her for many years before we did. I'm lucky.

The man who owns my company: Mr. H. He is sixty, slightly balding, getting heavyset like an ex-athlete, is terribly shy, terribly successful, has every material thing that ever was. He owns castles in Scotland, he owns jets, he owns electric garage-door openers, and when he wants wine he does not have to go to the store to shop. He's really only had his success, as they say, in the last twenty years. There's no one in town, or in the South, in the oil business who doesn't remember him when.

I work all the areas for Mr. H., all the ones he is interested or active in. Texas, Oklahoma, Kansas, north and south Louisiana (two entirely different worlds), Mississippi, Florida, south Alabama, north Alabama . . . Did you know that oil and gas have never been found in Georgia? I am told that the state is offering a million-dollar reward to the first person who can find some. It may be a tough chore.

I have fallen in love with the underground geology of north Alabama (and the area extending on over into the siltier portions of north Mississippi). The area up there that I work, where I spend all my spare time learning and theorizing, studying and mapping, musing, is shaped like a triangle, with the V pointing south, straddling the Mississippi-Alabama state line. At the basin's north end, the sands that hold the oil and gas—old buried beaches, Mississippian (Paleozoic) in age, two hundred and fifty million years old—are only about a thousand feet deep. You feel that you can almost reach down with your hands, reach in up to your elbow, and find the old histories, without the formality and expense of a drilling rig. And then, at the bottom of the V, as the old sea (what was then and still is now the Gulf of Mexico) retreated two hundred and fifty million years ago, in that deeper, narrower part of the basin, the sands, five and six thousand feet down, like plunging porpoises, sounding, headed back to the deep. And for now, that's the end of the basin, the deep southern part. Beyond that, it seems that the gulf ended, and Ocean, without sand, began.

There doesn't seem to be any more of the rolled-around sand grains lumped together with enough air space—porosity—to hold the thing I am looking for: decayed organisms, dead life, usually leaking out of tightly packed mudstones and into the clean, porous sands.

The pressure of the world sitting above these shales and mudstones and decayed organisms pushes down, and there is heat, and the hydrogens and carbons in the old organisms reassemble into oil or gas. The oil (or gas) starts to flow, to move again, forward if it can, and upward.

It always tries to climb higher than it is: moving, like a miner, through and between pinhead spots of porosity, trying to get up to the area of least pressure—back to the earth's surface, where it used to be. And like some video game, up it will continue to move, but slowly, picking its way, until the past earth—younger, higher rocks, but rock formations also buried—can trap it, and prevent the mutiny: blocking its upward progress with a fault, a blanket of impermeable shale, through which nothing can travel . . .

Not until time and ice and earthquakes and rains and rivers and such erode the mountains above this trap back down to the trap itself, can the oil or gas begin to move again. Or if a geologist can sense, as in a child's game, that there is something hiding, trapped, unseen beneath that shale, and can find the pocket where it is, and can poke an eight-inch-diameter hole into the trap, however shallow or deep it might be, then the oil or gas—always seeking the lesser pressure, the surface—is just about obliged to come out. It is as daring a rescue as ever there was.

This north Alabama area is called the Black Warrior Basin. There were a few wells drilled in it along obvious faults in the early 1900s—1907, for example—where oil and gas were actually coming up the planes of faults, all the way up to the surface, home free. They'd use that fuel for ridiculous purposes: snake oil, pitch and tar for boats, bathing cuts on livestock with it. Then there was another handful of wells drilled during both of the world wars, a few very shallow gas wells that were stumbled into—this great triangular buried sea, gulf, that they knew nothing about. They just cranked up and started drilling, *hoping*, true pioneers— and then in the sixties and early seventies, gas began to be worth something, oil too, and the basin has since become one of the ten hottest exploration areas in the country. It doesn't have the reserves of basins in Venezuela or the USSR, or if it does, they haven't been found yet. But it holds American oil and gas, and it is shallow, and people who want to work for themselves rather than for another person are especially able to work in this basin, since the wells are not as expensive to drill. Independents, these people are called. They account for a tremendous amount of the oil and gas found in the country. Not depending on anyone other than themselves. It's good that our country knows how and where to drill for a thing we need.

Some of the newer geologists call it the Warrior Basin. I always call it the Black Warrior Basin. That's its full name, and it is like writing Xmas for Christmas or USA for United States or calling the Black Forest in Germany the Forest, and I don't like it. Perhaps I am not compromising enough. But Black Warrior describes it better than Warrior, makes it darker and more buried and mysterious, and it's only the new people trying to shortcut in, playing catch-up, who try to cut off even the name. Most of the old guys who were around before it got hot, who helped make it hot, who knew about it then and were trying to keep it a secret, still call it by its full and right name. If something's bigger than you, then you have to show it respect, no matter how immobile it may seem to be, I think.

The ultimate moment in any well's life is when it is logged. For these wells in the Black Warrior Basin, logging is usually the instant of life or death, and I am the log doctor. I'm the first one to see, in the middle of the night, what we've got—whether I think the well we've drilled will be a dry hole or a producer.

A logging truck is ordered when we've drilled as deep as we are going to go (I look at the drilling samples and drill rates to decide where we are—have we passed through the

29

Carter sand? The Bangor lime? Are we far enough into the Bangor lime to shut down?), and the truck lowers a slender fifty-foot electronic tool into the 7⅞-inch-diameter hole of the well. The tool is attached to a spool of cable. The logging truck lowers the tool all the way to the bottom of the hole.

I am sitting inside the truck, watching a screen. The truck reels the tool back up out of the hole, slowly—more slowly than if you were reeling it in by hand—and foot by foot, the tool passes through all that dark mystery of time, emitting signals and picking up signals. I watch the tool's response to the formations it passes through on my screen, little green blips of radioactivity, and like an EKG, each blip indicates something. I know what to look for, and if I see the things I want to see—good high resistivity, good porosity, and good permeability, like the three cherries spinning into view at the end of a slot machine play—then we've got a well, a producer.

In the Black Warrior Basin, if even one of those variables—resistivity, porosity, or permeability of the rock formation—is not high enough, then we've got a dry hole, a duster, a dud, and my eyes, in the middle of the night in the high-pines backwoods of north Alabama, are the first to see that, too.

I love to log wells. I've logged a thousand, and I still find myself holding my breath when the tool first starts up out of the hole, when the electronic green lights begin to flicker and race.

No one has ever before seen what I am seeing.

Reverse history. Looking at the surface of the ground, up on top, in the present, possibly to get a hint of what it looks like below you, thousands of years below you, maybe a million or more. What went on Then is sometimes what is still (or again) going on Now.

Creekology is the worst term applied to it. Professors teach it at major schools, too, though they call it Surficial Geology, and give grades in it, take it very seriously.

There's a railroad that runs through the Black Warrior Basin from the small Mississippi town of Nettleton to Amory, to Quincy, to Greenwood Springs (and Wise Gap). On it goes, to Gattman, and beyond, across the state line, as railroads do. A solid one hundred years old, the path of this track goes, naturally, from town to town to town. Sometimes the line of these towns is straight; sometimes it circumscribes an odd sort of arc.

Why are these towns here?

Water supply, first. A creek, perhaps, or merely a mystically good place to drill water wells. Everyone wants a town in such a place. That's how they get started where they get started. And other things, sometimes: farmland. Good growing green things. Maybe a nice wall or bluff to the north or even south, backing up a valley.

Faults have to go somewhere. In a fault or an earthquake, one side of earth and rock is pushed up because of a difference in pressure far beneath the surface—a belch, if you want to anthropomorphize, or even Children Fighting —and one side drops. It's rarely a vertical, up-down drop; there's almost always some vector less than ninety degrees. Forty-five, sixty degrees—the plane of the fault varies.

Things often migrate out of the rock formations severed or disrupted by the fault, and move up the plane. There's a lot of rubble (air and dust and jumbled-up space), porosity, and lack of continuity where the earth has rifted.

Sometimes—more often than not, it would surprise you—these things from beneath the earth are able to escape up the fault, all the way from their past and up into the new, present world. A resurrection. Oil, sometimes, or gas, but most commonly water. There may be a lot of water, and it may be fresh and form a creek that follows, naturally, the trace of the fault along the earth's surface. And don't get me wrong: it's not like an open fairway, down this plane of the fault back to its origin, one thousand, two thousand, twelve thousand, maybe twenty thousand feet down; it's been covered with sediment, minerals, history and time, trees and civilizations, everything in the world, and pushed back together much in the way a broken bone or knitting scar heals. But it's still permeable enough—tiny capillaries between the particles of dust, clay, and sand—at times for a gas or fluid moving through these microscopic pores and pressured from beneath to work its way up. All the way up.

It's not as if a guppy or snapping turtle swimming around in the creek over the surface, where a fault emerges, could take a dive one day and decide to check out the Lower Cretaceous, or Jurassic, a couple of miles and many millions of years below.

It doesn't have to be a creek, where the fault decides (if it does) to exit the earth's presentness. It may just be where some mineral-rich water bubbled up, dried, then leached its minerals back down into the soil a few inches. It may be a band of richer soil. Or different vegetation. It may be a lot easier to drill a water well along the surface outcrop of the fault—there could be more water stored in the underground formations above (or below) the cut of the fault. Surely that's where towns sprung up, one hundred years ago in the Mississippi and Alabama wilderness—wherever it was easiest to drill water wells, or even along small creeks. And surely railroads and, later, highways are going to go from town to town to town, yes?

That is creekology, that is old-timey: looking at a map of the surface and saying, when you have no other information, or even when you do, "I bet that is what is going on down there."

For a fact, in Beaverton, Alabama (Beaverton Field: thirty-eight gas wells, one oil well, and only a small sackful of dry holes), there's an artesian well, a spring that bubbles clear, tasty water out into a roadside field in front of Shug's Café (fair coffee, good service). I'm not saying it's a fault that's bringing that good water to the top of the earth. I'm just saying it's in hot country, activity-wise. Things are going on. And it's a fault that traps the gas at Beaverton Field. That's been proven.

Where things are going on, you sometimes want to drill.

It is easy not to do things the right way. I do not know what the evolutionary aspect of this is, how it has survived. (Perhaps it will not: perhaps it is slowly disappearing.)

I'm a student of the earth, the woods, rivers and trees. I

know that there used to be ten million elk in the country (Maine to California, even in Texas). I know of the nitrogen cycle, the Krebs cycle. Feldspar is a six on Mohs' hardness scale. Plow your land straight across a hill, and it will wash away.

But how do you know when I tell you what's going on underground, or even above the ground, that it's true? And when I tell you where there's oil, how can you just believe it?

That is one of the things some people do: tell you they think there is oil in a prospect area to get your money, to drill, even when in their hearts they do not believe it for sure. They are just hoping. They are just wanting to drill, and maybe get lucky. Sometimes they do, and are. But that is not the right way, and very often they miss.

Art has been defined (sometimes) as being merely another word for selectivity.

Geology is, always, this: throwing out certain things and listening to others, the ones you think are true. Ernest Hemingway called the little (or big) machine in his body and soul that did this, with regard to his writing, a "shit detector." The old s.d. Throw out the bad; keep the good. It should buzz when you see or hear or write badly. Theoretically, everyone has one. Some are just abused, and others, ignored.

You can't accept everything. It's childish and dreamish. You'd be massacred.

You can't throw out everything. You'd be frozen, locked up in your indecision and fear. You'd never drill any dry holes, but then you'd never produce any oil, either.

You must learn, and listen, and watch, and practice, and do. Keep the machine oiled, and wince, horribly, when you see someone else make a mistake or, heaven forbid, you make one yourself.

Sometimes you need to listen to creekology and sometimes you do not. I cannot tell you how to tell the difference between the two times. I cannot make words tell you this: words are magical but this goes beyond even magic, into self, when to listen to creekology and when not to. It dies, here on paper, with type and ink. It is flat, and does not move, this string of sentences, and cannot make you know which instance is true and which is not.

Your heart is beating as you read this. You have a pulse. The words on this page are sitting flat on it and are not moving, will not move, even after you close the book.

The landlord and landlady's dog, Rusty, an orange chow, is the ugliest dog in the world. He was never attractive but I think he has since been hit by a car: his face these days seems to be turning in on itself, even more than the characteristics of the breed dictate. Contrasting with his ugliness is his great friendliness; he is always walking up to you and trying to put his head in your lap, as if he doesn't know he is hideous. I have clocked him at thirty-five miles an hour, uphill, running beside my truck when I leave: his tongue hangs out when he runs, and his eyes roll around in confusing and different directions. I *know* he has been hit by a car. But how can you help but like him? He likes himself, is happy with himself out here on the farm, so you are obliged to like him too.

Different states have different masters. In Texas, it is the Railroad Commission. In Mississippi and Alabama, it is the State Oil and Gas Board. Louisiana, the Department of Conservation. It is amusing to me that these regulatory agencies are good at what they do. I have been given tickets for driving 55 on the interstate. I have been taxed unfairly. I have been on the bad end of rules, it seems, more often than not and generally wince at the thought of them, along with waiting in lines, whistles, numbers (limits), and requirements.

But not so with the state rulings for playing the oil and gas game. It's not that I understand these rules better than any others; they're just fairer. Other rules are based on equality, or even distribution, or safety, or other worthy parameters, but in the end they all sacrifice, in little nooks and crannies—loopholes—fairness.

Oil and gas laws are brutally fair. Sometimes they sacrifice financial safety and allow individuals to lose all they have. But the key word is "allow." You might or might not have oil or gas beneath your land, but the rules that govern how you are going to get it out are consistent. Those rules have been made by oil and gas men themselves, not rule-makers.

I know you want an example right now.

In Mississippi, you can't complete (that is, produce from) a gas well if that well is less than two thousand feet from an adjacent well that produces in the same zone. Same for Alabama. I don't mean some people can and some can't. It doesn't matter if you're a congressman, or are someone

serving time, or if you're both of those things—if there's a gas well in your area, producing from a certain zone, and you want to drill a well and produce from that same zone, you had better be prepared to be two thousand feet away. (The rule for oil wells in Mississippi is five hundred feet.)

These rules keep other people from getting too close to your gas and oil; they don't let you get too close to other people's gas and oil.

You can get closer to another oil well than to another gas well because oil doesn't travel in the underground rock as fast or as readily as does gas. The oil is heavy and slow; the gas zips. That's why you have to stay two thousand feet away from gas wells, but only five hundred feet from heavy old oil.

A geologist, like any wanderer, explorer, or adventurer, is driven by just a little bit more volatile fuel. If a thing is capable of being felt, he seems to feel it stronger. Because the odds are always against geologists—because they're always more likely to drill a dry hole than not—they've developed a steely so-what ignore-it attitude toward catastrophe. They put dry holes behind them, repeatedly, immediately, and think only about good things. Childishness is as good a name for it as anything.

It is odd, but when I am out at the wells, staying up all night, the talk seems to move from food, to women, to drilling and geology, to good wells and bad ones. It is as if there is nothing else in the world, those nights out at the wells, waiting to see if oil has been discovered. Occasionally some hunting and fishing tales will creep in, but I classify those almost as food talk because as soon as they are finished, recipes are proffered.

I try to figure this out, and all I can come up with are the words "lust," "passion," and "gluttony." If you spend your life being hungry, missing more often than hitting (even the best geologists do this), then I guess those wants are sharpened in you. It seems that perhaps when you make your occupation out of such a life, spend your livelihood in the *pursuit* of something that is very hard to find, it puts a little additional stir in the other want-tos. It seems that those subjects, at least among able geologists, are the big three, with very little else fourth. And if Elizabeth, and the restaurants I visit, creep up too often in these notes, I am sorry. But not very.

I think it's chemical. Spending all my time *hunting* for a thing gets those other tastes and hungers alerted and receptive too, so that they, like the suspicion of where oil is, are felt keenly, sharply, like the high edge of a knife that has been honed for a long time.

I put air in the truck tires in the small town of Terry, on a bright day. It seemed like a major accomplishment: the sort of thing that could shift the pendulum, the momentum of the day, toward the good. Nothing bad could happen after that.

But more: we washed cars, her old dependable Chevrolet and my wickedly unreliable truck. If life and living are not like oil, then I do not know what is. When I have the fortitude to consider this similarity (though perhaps in ignoring the similarity lies strength), I know that it—both oil and living—will not last forever. I spray the hose over the hood of her car, in the sun, making her laugh. I know that she can never stop moving. But, unlike the way oil has been undervalued, I am also determined not to undervalue our relationship on days such as this, even years such as this, when there appears to be an excess of life.

An excess in the tanks, yes; in the jerry cans. But underground? How could anyone be fooled when stopping to consider it? We fall asleep hard, tired, holding each other. But underground, because she is Elizabeth, there is only so much. I do not give prices based on daily happinesses but, rather, on what is left in the world, in our lives. Each day, each time I see her, becomes increasingly valuable because some is gone, and I am happier. We may have twenty years left, or three months, or forty years, or more, but it is a finite unit, like the quantity of anything in the world. As I use the time up, I do not take it for granted. Rather, I try to be surprised at its continued presence and thankful that, at the surface anyway, its waning is not apparent.

We go to New Orleans. I order her a steak that could be used for an anchor on a battleship. The back of my hand on her cheek, touching it, that smoothness. Nothing lasts. Old seas are buried. A species of dinosaur may go extinct after only six million years. We will live to be eighty, with health failing after sixty: a fact. We drove to Arkansas one weekend, and ate fried fish at a lakeside restaurant at night. Moths batted at the screens and at the lights.

I used to have a company car, but too many things kept happening to it (them). The insurance company that covers all of Mr. H.'s businesses said that they had to cut me out of their coverage, like the soft part of an apple (they didn't say that), or the entire company's holdings would have to be released from the policy. I told Mr. H. that I guessed I understood. I was distressed but also, for some reason, proud in a sophomoric way. What Tom McGuane and the Catholic church call the hidden fun of "fomenting discord."

Once I sank one of my company cars in a swamp near the Red River. I had to swim out. It was frightening in a way I wouldn't have imagined; nasty, muddy water. I was driving home from the rig, having finished logging, going back out on the board road, which had flooded during the night: about half a foot of water stood above the boards. I had to go slowly and guess where the road was and wasn't by looking up at the lane cut through the woods, and hoping the boards stayed beneath me. A few of them were floating loose, pointing in all directions. What I didn't know was that there was a very deep drainage ditch on either side of the board road.

It all looked the same with six inches of water covering it: like a broad, shallow, muddy swamp. I drove slowly. I got off the board road and didn't know it. I was creeping along, water lapping just below the headlights. Suddenly there was a down-elevator feeling and brown water came rushing up over the hood and onto my windshield. I didn't even turn the ignition off; I scrambled out through the window and

43

swam two strokes back to the board road while the muddy waters closed in around the roof of my—Mr. H.'s—car. The car came to rest on the bottom, so that you could only vaguely see the opaqueness of the roof. The lights looked very dim down there, and then they faded entirely. Many bubbles.

I stood on the board road in water only slightly above my ankles and waited for someone from the rig to drive by who could pick me up and take me back in to have a little chat with the bulldozer operator. I was to learn much about bulldozers while out on wells for Mr. H., about the things they can and cannot do. For Christmas that year, Mr. H. gave me a twenty-foot length of chain. I think that sometimes I daydream while driving out to wells or coming back, but I can't remember. If I do, this is not a good thing, but how can you not help but think and muse while you are driving? I had to rent a car to drive home in, and I paid for it with muddy money and in wet jeans, wet tennis shoes, a wet shirt.

When you're going to or from a well, you can't dally. You can't let anything stop you. The earth and the well are bigger than you are but you have got to try to hold your own against them anyway. My clothes were bad smelling and swampy. When I got back out to the dark of the interstate I took all of them off and drove 70 mph with the windows down. It felt good, and fortunately for Mr. H. I was not pulled over. I stopped by my apartment, got dry clothes, and took the logs in to his château. We spread them out on the table in the kitchen, discussed the different formations, and decided to run pipe on it, to attempt to make a well. I forgot to tell him about the car. Everything is so much smaller than the act of finding oil that it must be like belonging to a cult, I would think. It takes you over. It gets in you. You feel as if it is *you* and not the oil or gas that is trapped down there, being pressured down. You want to find it, and have it come rushing up toward the surface. Like a diver, deep below, looking up at the bright of sky: rising.

I don't know if I've told you about leases or not. I'm in Houston now; late November, morning, coffee, gray sky and fog. Bare feet pleasantly, almost luxuriantly chilled; relatives/post-Thanksgiving time of year. I went deer hunting, used the last of my vacation time to go out near Austin with cousins, grandfathers, uncle, father, etc. Elizabeth flew in yesterday to visit.

She is in her room doing a workout (much dancing about and flailing of arms; sounds of thuds and muted athletic thumps issue from behind her closed door). Ten-year-old brother B.J. is silently, cautiously awed.

Sigh. Leases cannot be avoided.

I really, really like talking about the cafés where you stop, the people you meet on and off the rig, and even the wet, sweet blackness of the oil saturated in those frosty sand grains that look like sugar, and how when you cut a core and try to hold a piece of it in your hand it is sometimes too slick with mud and oil to get a good grip.

Plus, I really don't even know about leases. Okay; I will wade through it. I will leave out some of what I know and a lot of what I should know but don't.

This is really basic. You can't just go out and drill an oil well. You have to have control of the mineral rights; you have to have what they call a lease.

The man who owns the land where you want to drill does not necessarily own the rights to the minerals beneath the land. They may still belong, in entirety or in part, to some old bearded pioneer who lives out in California in a tent with no address.

Many times I have been mapping—soft-leaded HB pencil skating down and around ravines, across wooded ridges, curling lovingly around hillsides, past charming little white frame-house churches (all this unbeknownst to us, for we are below the earth), single-minded, tracking, stalking a single formation, a single dark buried streak of rock that behaves not at all like the rivers, valleys, churches above it—and I've stumbled onto oil, into a vault, a hidden place where I believe it might be hidden. (An "accumulation" is the word geologists use in papers and at conventions.) And we geologists get excited, walk with our arms flared away from our sides a little more than is necessary, polish up the area on a base map, and with that good springy feeling in our calves go into the exploration manager's office.

If the exploration manager likes it, he takes it into Mr. H.'s office (sometimes that day; sometimes a week, a year, two years later). The timing depends on attractiveness of prospect, other commitments, local prices, and market demand for the hydrocarbons. Even if we do find gas, sometimes there's a glut and we can't sell it.

But anyway, the point of this is, if the prospect is accepted, or deemed worth looking into further, we, the land department, have to discover who owns mineral rights to the individual tracts of land that lie over this new buried Spindletop.

It's wildly frustrating, usually. There are a thousand and ten items to be done before your well is drilled.

I am thinking about it now, and perhaps it is a mixed blessing, because it does allow for anticipation.

But on the really sure prospects—the ones where you can actually feel the oil straining to get out—it's infuriating. Other oil companies may already have the mineral rights, or may be simultaneously competing for them. Occasionally there are disputes in which several oil companies each believe they have the lease. Oh, wily landowner! Or the landowner (rather, mineral rights owner) simply might not want to lease; it's certainly his or her prerogative. Some people like being poor; that's fine.

So anyway, once you "find" oil—on the map, that is—you're usually in for a wait. And rarely will you be able to, or even want to, obtain *all* of the mineral rights in a drilling unit. Say it's a gas prospect (you *think* it'll be gas, anyway; let's suppose that most of the other production in that area is gas), and the local spacing rules (defined by the state's Oil and Gas Board, or, again, the Railroad Commission in Texas, the Department of Conservation in Louisiana) allow for the creation of 320-acre drilling units. These units may consist of half of a section: east half or west half ("stand-up" units), or north half or south half ("lay-down" units). Well, you might have been able to obtain rights to only 160 acres, one half, of the unit you want to drill first. And oil company B down the road might have, from leases taken sometime in the past, say, 58 acres in that unit. And an independent speculator, Joe Horne, might have 6 acres. And maybe a local wildcatter kind of likes the area on general principle and has another 37 . . . At any rate, if you have 50 percent of the acreage in a unit, this means that by and large you'll be paying only 50 percent of the drilling costs. So instead of $100,000 to drill a well and analyze it, it'll cost you just $50,000. Of course, when oil and gas start coming out of the ground, you'll receive only 50 percent of the profits, after taxes and royalties and such. But it is still not a bad idea to have the leasehold interests spread out.

You hope and prefer that they are spread out among knowledgeable and competent companies or individuals, but you have no control over this; you can't choose your neighbors. There's nothing quite like drilling a well with a partner who won't or can't pay, or makes trouble, or is rude, irresponsible, or the like.

———————

And there are strategies within strategies for leasing. I swear you can tell a land man by the way he walks.

It's not exactly like being a used-car salesman, but it's close. The men who negotiate leases offer fair prices, they're not crooks or anything, it's just some funny aura that labels them. I don't know how to explain it, unless it is plain old muscled-geologists' envy at their glibness and dexterity with clauses and documents, an ability to hold a sincere, entertaining conversation with tobacco farmers and tycoons alike. They know how to buy a man's mineral rights and get the best deal for both sides: oil income and money for both parties.

I suppose it's disloyal, and some geologists might quickly disclaim it if you ask them, but this is how it is. I swear, there's a perverse sort of chuckle-pleasure that ripples through the geological department on that rare occasion when an unruly or stubborn landowner really hangs tough and sticks it to the land man, leases him the land, all right—yeah, we get to drill the well, all right—but out-glibs the land man and rides him down to the ground and makes him take a deal that is less, much less advantageous than what the land man had set his sights on.

You can't be too serious. Not all the time. If you can't grin at the humor of being beaten occasionally, you'll be horribly hard to live with in victory.

If I am making getting a well ready to drill sound almost like a game, I will not stammer or back over myself in trying to dissuade you that this is not so. Not at all.

G irl in black sports car drives up matter-of-factly. I am sitting by the phone, on a bench with a farm kitten, waiting for a phone call from the well. I am to meet Elizabeth and Flynt out here, so this is known as double late. But it's a wonderful day; I'd rather, I think, be doing nothing other than sitting on this bench.

Lazy December, Saturday. Eleven A.M.? Maybe closer to noon? The stables. I like the way that sounds. I like the fact that I am at The Stables. Two wells up north, late tonight and tomorrow, and no Christmas party (I wasn't wild about going anyway), but that is all later. The air is delicious, and the girl—old jeans, tennis shoes, a nondescript old button-down shirt, with a black suede coat making her look a little like a ringmaster—has led a young, big gray horse out into the corral. She's handling him well: chasing him around and around, playfully. Though his eyes are only partly mock-wide with alarm, a lot of it is real.

His exercise. One year at Thanksgiving (before I joined the oil business and when I was home for the holiday) my uncle was cutting his steak (this was Texas: eighty degrees outside) and talking about horses. He shook his head world-wisely and said cynically that the bigger an animal was, the smaller its brain, and as I had been working out with weights a lot then and weighed about thirty pretty hard pounds more than I had when they had last seen me, the whole family looked immediately and significantly at me, and then we laughed. Families are good in that they do not let anyone else tell them how to think or what to be. I like the way they stay together.

The driller called and said they had only about seventy feet left to drill before reaching T.D. (total depth; a good sound to any geologist's ears—tee-dee, tee-dee), but it would be in the Tuscumbia limestone, of course, and slow going. Elizabeth and Flynt drove up, and we watched the girl and the gray horse and then Elizabeth rode Taco, Flynt's horse, down through the fields. Flynt is Elizabeth's brother and is thirteen, the size of his shoe.

Late that night, about midnight, maybe a little after, headed north up the Trace, going out on the wells, I see the biggest buck I've ever seen in my life cantering across barely in front of me, so that I have to brake. He continues on into the brush. I will almost certainly never again see a deer that large.

I think about the size of his antlers—nearly two feet tall, dream quality at that time of night, and so unnaturally large—for the rest of the drive. Thoughts freeze, lock in, and float, hanging, while driving between midnight and five A.M., especially on the Trace. No traffic out. Just deer and owls. Your headlights catch them swooping off the road.

I log the two wells that morning and afternoon, and drive back around dusk. Drop the logs off at Mr. R.'s, the exploration manager's, house. Continue south to the farm, think not a bit about oil for thirty minutes—release—and go to sleep immediately after eating three gingersnaps and a box of Cracker Jacks. Wake when the sun comes up the next morning and drive back in to work. Mr. R.'s house is in a new subdivision surrounded by all these other houses and really looks kind of funny, out of place, for it is California adobe in the Old South, with Spanish moss, but he likes those sorts of houses. Until I go out on my own, he is my boss. Sort of. God, I drive a lot. You can't ever be late, and it's 250 miles. And you're always running behind schedule. Sometimes day, as opposed to night, loses significance, and also you feel like you're being washed down a mad stream somewhere, or a river in a flood, and are missing things.

You want to catch your breath, but there is another well.

And another. And another. But it is fun. You go beyond your limits, chasing it. Fatigue becomes the currency with which you pay.

It makes sense, though. It is energy, after all, that you are looking for: buried.

It smells so good, so clean out here. All geologists are hyperbolic, but this is the most wonderful, most harbored road in existence. Not one mile away, Mr. H. (big boss, real boss) is drilling the ATIC 33-2, in Escambia County, south Alabama. The road is patchy black and gray asphalt, running through subtropical ferns, hardwoods, and standing black swamp water. The smell of rich life, stagnant in spots, and the noise of crickets. Some are chirping slowly, some fast. Obviously down in a swale. Someday this road I'm walking will show up on a geologist's log as a coal seam. A deep purple dusk now; the clang of pipe tongs not so far off, ahead and to my left.

Definitely a temporary sound, a brief, frenetic energy. Much too intense, compared with the subtle strength of the ferns and oaks, dragonflies and jays, and slow crickets, to be sustained. Mourning dove in the early night. I have no flashlight. It sounds sad, sweet. Slow. Demanding nothing.

It is peaceful, walking here. Up a grade now: my muscles feel wonderful. I am glad for the weightlifting: I do not doubt the hours spent. A few pines begin to show up in the hilly stretch. Skyline ahead; a bird flutters in the near woods. Elizabeth and I went to a movie once in Vermont, a late, crisp night, about ten; both of us excited. We went bowling afterward.

These pines are young, seven years perhaps, and have black scorch up to about five feet: grass fire. Grass fires weed out all the little stuff, weak vines and useless tendrils and thin grass, and let the pines grow, go on.

Road goes to dirt now, a pleasant surprise. Red clay and

grit: pine country. We've been talking about the marriage thing. I think I might want to try, am nearly ready. You know: children, the slam of screen door, etc. It frightens her to death.

The dreams of the wildcatter, the dreams of oil not seen, future oil. What is it in a person that makes dreams of the future so very much dearer and more valued than what is the present, or what was? Maybe this is not even a thing exclusive to humans: the other side of the fence.

The smell of these woods, the peace of nightfall, a half moon, my life. What is it in me that makes me literally have to stop and pause, sometimes stunned by their actualness, their beauty?

These noises, and the present: they are so real, so whole. It seems that there has never been anything more real than that katydid, rhythmic and chirping, high in a tree and to my right, which I cannot see as it is near-total darkness now.

The anticipation of drilling—it could be the desperate hope of making these things of the present stay real. Test the future, and establish that it too is real, to guarantee that these things now real will continue as such into the future. Sure, you've found oil once, today. But can you find it again? And after that? Only an appreciation of the present, what you have, can spur you on to such a violent extent to keep testing the future.

The desperate need to keep checking out the future. Maybe that's the urge to drill ahead, drill deeper, drill updip, keep going. If this need was disenchantment, rather than appreciation of the present, I would never have the courage to find out the real truth. I'd hope, and create wishes, but wouldn't dare look.

Except that here a dry hole doesn't mean that the future's not real, that oil, or whatever it is you are looking for, was missed, or isn't there. It only means that the right place was not tested. It must mean that. No option. That is what it has to mean.

It's a powerful secret, to be walking around as free as a whistle and realize you know how to find many, many thousands of barrels of oil—an average well can easily make a hundred thousand barrels—right beneath the ground you may be walking across or drilling on. I will try to explain the dizzy feeling you get when you are first on to that secret: in your office, you are leaning over your desk, looking at parts and pieces, clues, and then you see it. It really does feel as if the whole earth is swaying, and that that oil locked in, trapped beneath the rocks down there, is all that is constant.

In the winter in Utah, where I went to school, everyone would get restless, feeling trapped and burdened.

Not me. I loved it. It was so cold. I would put salt on cheese and eat it. Movies were fun. Winter never lasted long enough. No one was out. Walks were like you were the president of snow. You could make the snow your friend or your enemy. It is the same with loneliness. You can feel trapped, and suddenly, one day, you turn the corner, and you are the president of yourself. You fix a sandwich. A bird calls. There is no feeling in the world like being strong. If you can take being weak long enough, it in itself will make you strong.

We are all going to run out of oil very soon. Our country sooner than others, to be sure, but all of us down to nothing, Venezuela and Arabia and China too, and it will be very interesting to see how we handle it.

Broiled soft-shell crab. Key lime pie. Tomatoes. Smackover. Norphlet. Reklaw. Carrizo.

Girls with big chests. White silk blouses at the receptionist's desk. Long hair that hangs straight. Calves that have been to aerobics class. A hand holding a glass of wine.

Shish kebab. Mediterranean beef. Melon balls. Paluxy. Rodessa. Tokio. Catahoula. Pine Island.

Big eyes. Girls who smile. Girls who frown and set their jaw. Girls who drive fast. Girls who drive old cars.

Cheddar cheese. Rack of lamb . . .

I will try not to sling terms around without telling you what they are. The tool pusher—more often called just pusher—doesn't push anything. He is in charge of the drilling crew: roughnecks and roustabouts. The driller is in charge of the pusher. The driller has got to be a mean sunuvabitch. The pusher has to be the second meanest. The crews, in the aggregate, are called tours (pronounced *towers*, and I don't know why). There is an evening tour (twelve hours) and a morning tour (twelve hours) on most inland drilling operations.

Everyone has to wear helmets, hard hats.

Whenever there's lightning you stop.

No smoking.

Sometimes the pipe will get stuck. You can't go in any deeper but also you can't come back out. When this happens it is the driller's responsibility, even though it might have been the pusher's fault or inattention. It can be a million-dollar problem, and it is the chief thing that makes drillers so mean: fear of getting stuck. Of having to abandon the hole, junk it, and walk off. Of not allowing the geologist even to find out whether anything was down there or not. Never knowing. Having to second-guess, forever.

You don't have to drill straight down, either. You can drill directional holes—also called crooked holes, slant holes—and they are remarkable displays of technology. Say you suspect there's oil under One Shell Plaza at a depth of ten thousand feet. You can back up to the nearest vacant lot, move your rig in, tell the engineers where your target is, and they'll put a directional bit on the end of the pipe, the drill string. They'll drill at an angle, taking surveys as they go and correcting accordingly, until they drift across the target, and then they'll drop straight down. It is like magic.

Anything can be done in the oil field. If you think there's oil somewhere, and have the dollars and desire to find out, it will be done. There is no such thing as "I can't." It is wonderful.

We go way back. Long before I even knew her—back when I was still in college, studying rocks and other things that I thought then were hard to understand—I would do stuff with her brother Flynt. Take him camping, take him for plane rides, for late-night suppers at the pancake house. Sometimes I'd see her around.

Our first date was this: she went to Riverside Park with me, on lunch break, for a bacon-lettuce-and-tomato sandwich. I'd talked to her some, briefly, and knew that she liked them. She'd been eating one once when I was over doing something with Flynt. I also knew that she liked to go to Riverside Park a lot. I'd seen her out there often when I was running, lying on a blanket reading a book. Riverside has lots of open grassy space. I knew it would be a good place to escape to from downtown for a summer lunch. I put my microwave and extension cord in the trunk of my car. I bought bacon and wheat bread and fresh romaine lettuce and mayonnaise, and kept them in my refrigerator at work, and waited for her to wander by my office and mention lunch, as she sometimes did. Just friends, you know. And I'd say, Sure, call me . . . But it never seemed to materialize.

I had to keep replenishing the lettuce. I'd keep it fresh. When she wouldn't come by, I'd fix a salad or something, and buy some new the next day. But then one day she came by and said it again, fleetingly, flirtingly, beautifully, sunglass stems in her teeth, tan and long-haired: "We'll have to have lunch sometime."

She was already drifting out, but I said, "Yeah, that'd be fun." I said it with real interest. Back then I had trouble making the words convey the excitement I had for things, so it was a real effort. I think I was afraid, for reasons I've long since forgotten, that other people might not share my excitement. I didn't dare show Mr. H. any of my prospects for two years for this same reason. "In fact, I'm going over to the park today for a bacon-lettuce-and-tomato. Would you like to go?"

You and I both know she wouldn't have gone if I hadn't dropped that part about the BLT, but that is beside the point, because she did go. Yes, she went. I opened the car door for her. It was hot. Out at the park I plugged the microwave, without even having to take it out of the trunk, into the outlet beneath a light pole beside the tennis courts. I fried the bacon in the microwave, spread the mayo, and washed the lettuce under the water faucet in the shrubbery by the courthouse. She grinned the whole time, looking at me from behind her sunglasses, and we ate those sandwiches.

She didn't go out with me for another solid year and a half after that. I swear it. An odd courtship.

Now we eat lobster in Acapulco. I like to tell her this. It sounds rich. We went to Acapulco once, on the ninety-nine-dollar special out of Houston. The peso was gutted, and we did have lobster. Three times in the same day. Gluttons (oil-men).

Bogue Chitto. Present year, 1985. Supposed to log one of G. Green/Chas. O'Neill's (Laurel Operating Company) wells, one of those directionally drilled wells out in the Intracoastal Waterway, so I passed up a whitewater trip to north Alabama. I mean, I had the gear loaded. The canoe was on its rack atop the truck. Even groceries; I had gone that far. The well wouldn't be ready to log until late Sunday night, so I was going to drive up north, paddle Saturday and Sunday, and then come back down through Jackson and on to south Louisiana (I like to drive; I also like to paddle) in time for the logging that night. But the drilling speeded up. They kept drilling sand instead of shale (sand drills terribly fast, because it is much more porous, a lot faster than does tight old impermeable shale), and as a result were expecting to log sometime Saturday.

This is a long story for a little explanation. I skipped going to north Alabama. But Saturday, when I phoned the rig, I discovered they'd gotten stuck during the night. This happens often on directional holes. They were coming out with the pipe to put a newer, sharper drill bit on, a "bit trip." It is terribly hard on a bit to drill that abrasive silicon sand, even if it does drill *fast*, and they'd gotten stuck.

So. I *had* to go paddling somewhere. It was one of those weekend urges, necessities that border on primal importance. Elizabeth and I went to the Bogue Chitto, even though it was storming, because we *had* to go. We cooked steaks in the tent on a little stove, and about ten the rain stopped and the moon came out between the clouds and shone on the water. We camped on a small gravel spit of an island, with driftwood and a few large trees. We sat outside for a while and wished we had coffee. The night air had spring and dampness and freshness in it.

In the morning, the weather was pale blue and mild, but the river was a torrent. We were much excited by this. We were to paddle down to a bridge about eight miles away and then hitchhike back to the truck at our put-in. We had the tent, sleeping bags, ice chest, everything in the canoe, and there were haystacks on the normally tame Bogue Chitto. Rapids, falls, plunges: just like the rivers up in Alabama and the Carolinas that we usually had to drive for hours to get to. It was a very pleasant surprise, a gift, to be given a whitewater river on this day. The unexpected presence of adrenaline on a calm and pretty Sunday morning. We did not take it for granted. We paddled well and did not turn over, although our hearts raced in several places. That was the best.

But there were slow stretches too, and we didn't take them for granted, either. Marl (folds and textures of twisted clay formations) undercut ledges of the riverbank along some sections. Gravel on the sandbars; map turtles diving off the logs. Geology; biology. Near the end, a slight mist moved in and started falling, enough to make the river seem spooky, surreal. I remember thinking: By and large I am happy, and not entirely sure why, but I want someday to get back to the mountains—Utah, Montana, and Denver, Colorado. I don't know if she'll come with me or not. I guess it doesn't matter today.

The farm in Terry has horses now. We drove to Texas over the weekend and my parents gave me one of their horses from Goliad, the irreconcilably wild Red Mare. She is too mean for a fuller name. It took us five hours to load her into the trailer. A tough horse.

"I don't especially care if I ever see her again," my father sang as we drove off. He was smiling, his eyes were wide and bright. Red Mare had thrown him six times the day before. She refused to be saddled like no horse we'd ever seen before. The people at the stables who do that sort of thing for a living refused to have anything to do with her. One of them had had his ribs and a hand broken by her, when he tried to break her. I like her very much, and the line of her jaw. She came out of good stock, too good, Skippa Star, and was too valuable to turn loose on an island or sell to a circus.

"Good-bye!" my father cried. He ran alongside the trailer as we pulled away and slapped the sides of it. "Good-bye! Good-bye!"

Red screamed in her anger and kicked, not at the slaps, but at being separated from the other horses. She was rocking the trailer, which was a little sad. I am told that all horses are like that. But her new beau, Taco, was waiting for her in Mississippi. We'd moved Taco down to the farm. Taco, too, had screamed and fought, leaving his stables. The guilt, after moving him, lasted for about a week, and felt like a wet blanket draped over me. I was pretty quiet, all the way back to Terry, with Red. I could never be a rank wildcatter, I don't think, drilling somewhere I wasn't *positive* there was oil. Disappointing the investor—that would make me feel guilty too. I hoped that Red and Taco would be interested in each other, gelding that he was.

Whenever I stopped for gas I went back and peered through the slats at Red. Her eyes were wet and huge in the darkness. Her metal water bucket was trampled and flattened. I didn't want her to be unhappy, but I didn't have much say in it, and it made me feel helpless. Plus that damned second-guessing. Wondering if maybe I should have left her in Texas.

It is cold tonight. Occasionally it gets so cold out in the oil fields that the gas won't flow. Sometimes, depending on the gas reservoir from which you are producing, there are little water droplets mixed in with the gas. Once inside the pipeline, they can freeze and entirely constrict the flow lines.

The flow lines are the relatively fragile little lines that go from the wellhead to the tanks, if it's an oil well, or, if it's gas, to the larger, sturdier pipelines of the gas company. The pipelines are sometimes under higher pressure than your gas well, if you have a little one, so you have to rent a compressor to build up the gas you have to pipeline pressure. "Bucking line pressure" is what they call it, trying to get your flow-line gas into the big pipeline when there is a problem. It's so much like life, this business of oil, at least what used to be life.

Besides cold weather, hurricanes cause difficulties. Hurricanes will lift those little (2⅞-inch-diameter) flow lines and throw them about, snap them, curl them like pork skins, ruin your uninsured life's work, make you cold at night when there is no gas. You can imagine the fear the gas well owners both big and small live with out in the Gulf. They must insure their wells, always, both those producing and those still drilling.

Like I said, it is cold tonight. I light my heaters. They hiss. It is childish to imagine a dinosaur making this sound. A brontosaurus coiled up in my butane tank; a triceratops's hydrogens and carbons swirling through my little copper pipes, igniting upon his grand and yet anticlimactic reappearance into life, my living room, *60 Minutes* flickering in the other room . . . Hissing, roaring in old dinosaur language . . .

Lost souls? I don't know. I'm sleepy.

I wonder: if I worked for myself and drilled my own wells, would I be able to get to sleep as easily as I am going to right now?

No matter. I have to go on out there, into the bigness, the scary part where declining pressures, water droplets in flow lines, and hurricanes are. Where it will affect me directly.

There is no choice.

I have much too great a passion for it these days to want to be doing it through someone else: a boss, an owner, or any other individual, no matter how benevolent.

It is like getting married. There's nonsense and strife and other things not perfect to put up with, too, along with the part that makes your heart lift and your life move, but the way it feels between knowing the oil is down there, under mere earth and rock, and knowing what to do about it . . .

I'm not sleepy anymore. I am going to have to read for a while now.

Winter. I have a craving for potatoes: I fry them, bake them, boil and sauté them. The horses stand in the middle of the field, broadside to the sun. Reasons. You can find oil if you remember to look at all the reasons; things below the earth, and even those going on above it.

This is how you get paid on an oil well.

Before taxes, oil is worth, say, $28 a barrel. You have a well on your land, a forty-acre ranch. But the well might have been drilled as an eighty-acre unit. This means there's another forty acres of the unit that you don't control: your neighbor, in whatever direction the unit happens to lie, is your partner in hope. Units are good things, by the way, because once a well has a unit assigned to it, it is subject to rules, such as the one that says no one else can drill on your unit. It locks in that eighty acres of land, and the oil that lies beneath it (in theory), for the person who owns the mineral rights to that eighty acres. Which, for simplicity's sake, we'll say happens to be the landowner. (Occasionally a landowner will do something drastic, in response to dire, sudden financial need, and sell his mineral rights to someone yet retain ownership of the land, or even just sell a troublesome fraction—a fourteenth, a seventy-ninth—but that's not too terribly common. But it doesn't have to end there. The landowner might divide up his mineral rights one acre at a time: an acre to a cousin living in Santa Monica, California; another acre to a widowed sister's aunt who, the last thing he heard, was running a trapline in New Zealand . . .)

But back to the sample case. You have rights to forty of the eighty acres in your unit. The well is drilled and completed, flowing a hundred barrels of oil per day (100 BOPD). So. You have what is called a half interest in the unit (40/80 acres). Multiply this half by $28 per barrel by 100 BOPD and that well will earn $1,400 per day.

Now you are royalty.

This is where your past returns.

The oil company drilled that well. They hired a big staff, maybe had some of those people on the payroll for twenty or thirty years. Every day, in and out, lots of days. Lots of overhead, too. Big things like compressors and rig rentals (ever try to rent a machine that will drill a hole a mile deep into rock?) and blowout insurance and bulldozers and copying machines; little things like coffee for the coffee machine. They're a rolling, blowing, going organization with all the usual inefficiencies, and let's just say for a round figure that it cost half a million dollars to buy the leases to the mineral rights and drill and complete that well. (Completing a well is like getting serious about someone, I would think; you haven't *seen* expenses till you've drilled it and are trying to coax the oil and gas *out*. There are a thousand little fragile, procedural, technical-engineering things you must do, all kinds of equilibriums and balancings and chemical stimulations, and they're all very, very expensive. Plus, you've got to buy a nice ring: three thousand, five thousand, or fifteen thousand—or however deep your zone is—feet of heavy steel pipe.)

So those dollars have been spent. If by some dinosauric quirk of temper there happened not to be oil beneath that eight-inch-diameter hole in the ground, and maybe not even any of the sand they're looking for—maybe the old river channel went left around that Jurassic tree instead of right—then that's a dead solid five hundred thou that is gone—swoosh, the earth sucks it up, a little bit of science and absolutely no oil, nil gas.

74

What I am leading up to here is a defense of the reasons that the oil company, or the individual who drills the well, gets anywhere from three fourths to seven eighths of that $1,400 per day. The amount is agreed on when the lease is signed. And why nobody minds? Symbiosis, the biologists call it: good for both sides.

The oil company bought the lease, the mineral rights, from you. A deal's a deal. They do the work and spend the money and give you, *after* they've paid you for the lease, 10, or 15, or 25 percent, or whatever. Big money, considering what you have (or, rather, have not) put into it. There's an awful lot of oil and gas wells supplementing an awful lot of little old ladies' Social Security checks. Lots of farmers now able to afford tractors, combines, etc., with which to continue their hobby.

So you've got the money from leasing your land—ten, twenty-five, seventy-five, a hundred, a thousand dollars an acre; it depends, of course, on how hot a trend you're in, if any, and what kinds of wells, if any, are being made in your area. You've got, say, one quarter of that $1,400—$350— coming in every day (actually, accounts are paid monthly, so that would be $10,500 per month), and other than what the federal government decides to take from you (say, half), the rest of it is still yours. They take it from the oil companies too. Makes us not want to drill sometimes, a lot of times; makes an awful lot of areas economically marginal or sub-marginal. Sometimes a well will go for twenty years once completed. In his pleasure, a landowner may throw a barbecue for the geologists. Those are fun. There will be a pig roasting. And an ice chest of beer. And corn. Maybe he'll let you go hunting with him on his land. It's sort of like marrying his daughter, finding oil on his land.

Gas.

Gas is neat, too. You can't see the gas, can't climb up to the top of the tank battery and peer down into it, peer at its hot black sticky sweetness, but it smells good, and is more impressive to me than oil, in its own strong way.

It hisses.

It really does scare me still, makes me much more conscious of the power of the earth and her history and even beyond that, of physics and the universe—it gives the earth more character, makes her power three-dimensional, when that wellhead is opened up and the high-pressured gas whooshes out of the pipe and into the air. Gas from the earth's insides herself: pure energy (you caught it!), raw force, the real thing. Believe me, the real thing. It is not Hollywood making that dry roaring; it is something that is real and that you might not have dreamed was down there.

It's not like anything you've ever heard. I could say it's

like a jet, like a comet, but it's not like anything else. You just need to go out to the field and get someone to open up one pipe for you for a couple of seconds. On an average well—one that makes, say, a million cubic feet of gas per day (1,000 MCFGPD, the M representing a thousand)—a five- or six-second demonstration of the earth's power and angry internal rumblings, angry desire to escape, will cost about seventeen cents. Yes, gas is cheap.

Y ou'll try to drill a dry hole on purpose, a shallow little well for disposal of saltwater that's being produced along with the oil in another well, and you'll punch through some accidental mystery sand that's got maybe a couple of years' worth of gas in it. You'll do the wrong thing and it'll be the right move. Perhaps you'll be downthrown on a fault when you're trying to be upthrown, and still you'll make a well. And all you are supposed to do when this happens—this is your obligation, in fact—is to just squint your eyes into the sun and nod slowly like some fat guru and don't tell anyone you didn't know what you were up to, instead just squint at that sun and consciously breathe in every stolen second of it, because the other ninety-nine times the other kind of luck is going to pick up a stick and knock you into the next county with it. You got lucky this time, but the unexpectedly unfortunate will happen to you too, far beyond logic or reason, and if pressed for a reason, all I can guess is that it is some negative energy or power of the earth that toyfully slaps at the lives of men for even daring to try to touch her. All in good ironic humor, of course.

As if a dry hole or any small or large amount of dollar bills mattered to the earth, and to the ice caps on mountains, and to forests growing on the north side of mountain ranges.

If you learn all you can, stay a jump ahead of the disasters, dodge the bad luck, and actually *think* like a strong, buckling, folding, changing, oil-trapping earth, then maybe the odds of that good luck and bad luck, science aside, will be only six-to-five against you. But if you go running around being curt and obnoxious about it, excessively optimistic and wired on luck, punching holes in the ground based on an itch . . . it is then that you destroy the thing in you that will let you *find* oil, going out and *getting* it, rather than having it bump into you. And the bump way—it won't happen enough; you'd have to live to be a million to make it worthwhile. It's not the way to do it. My job is like living.

This is not to preclude the itches. Just add them to all the other factors and variables. Don't rely on them so much that you find yourself using nothing of the other.

And this is not to say that there is no luck, either, because surely there is. Sometimes you will miss your objective completely and instead accidentally discover some new stray sand with oil or gas in it that no one's ever seen in the area before—like the Benton sand in Fayette County, Alabama; it's found only in a couple of sections. It's just that it doesn't happen as often as the other kind of luck, the bad. Just remember to be real conscious of it and make it taste good while you have it, when you do stumble into the good.

This is one of the ways Elizabeth is different from the oil business:

She's lucky. She's lucky. She's lucky.

We are driving out in the desert in New Mexico, not far from some nameless little ghost town through which neither of us has been before. It is a three-day weekend, and we are a long way from home. Headed to the mountains for a few hours—eighteen, twenty, twenty-one if we cut it close.

Anyway, out of the nothing, she says, "I wish there was a Y or something around here, just so we could run for a while and then have a place to shower." I really am only sort of half listening, and then I turn right and go on up the road a few blocks, past some gas stations and stuff, and then out of the town: we're back in the desert. It is a little sad to be leaving even the ghostly semblance of a town so quickly. There is one last small dirt road branching off our weeded little two-lane state highway. As we draw near we see that there is a hand-painted sign about the size of a license plate, stuck on a garden stake and leaning weakly to one side. It

says YMCA, and there is a faded arrow pointing out to more desert.

We look at each other with two different expressions. She is delighted, a joyous child, this event no different from any other single event of her life—Fortune's child. Chocolate ice cream for breakfast, of course, please. I myself am somewhat and openly frightened. A little.

Still, a run in the desert will be nice.

We stop. We lock the truck, change, and run. The facilities at the Y are nice, just adequate, and there's not a tremendous amount of water, either hot or cold, but their sheer existence elevates them to luxury.

Don't you want to know what my office looks like?

My wood desk groans. Clockwise, from nine o'clock position: oil and gas journals and *Southeastern Oil Review*s, a couple of months' worth. Eleven: drilling reports (pink, Texas; white, elsewhere). Eleven-thirty: logs to be marked, made sense of, explained, and mapped. Twelve-thirty: more logs, more clues. One o'clock: phone book, Kleenex box. Two o'clock: field files—some Texas ones right now, but they vary. It depends on what is happening, and where, at any precise moment. Sometimes thoughts for the future are taken into consideration, but usually there's no time for that.

Three o'clock: still more logs, and I can see a tuft of a sidewall core analysis from some who-knows-where well sticking out. Truly, some stuff does get forgotten over the course of a battle of daily events, all of them crises at their

81

own moment. I tug on it, out of curiosity. It's from Nueces County, Cotton Field. Oil shows up and down, goodness: 6,854 feet through 6,864 feet analyzed as oil productive with relatively low water saturations—all around 40 percent— and all the samples in that interval had bright yellow fluorescence. Perms were okay, ranging from 36 to 119 millidarcys, porosities in the midtwenties . . .

It is past five o'clock, Monday. Drear. I can hear my watch ticking. This is the point where I can *hide* behind my desk, like a gunner behind his sandbags. I am going to write awhile and then go lift, then go down to the farm, where I've left the heaters going. (Twenty degrees out there: forgive my softness in my old age, please.) Hooray for butane. I love walking into my warm farmhouse. Yes, I could use a wood burner, but I don't want one.

My office. Beyond the desk, my microwave, over on the credenza that looks back at me from the wall. On dreary or cold days, I like to sull up here at lunch, and in the evenings too, and write with the door closed. I'm up here more than I am down at the farm, and besides, at the farm, I always have more time. Up here is where I need the microwave the most; up here is where I have it. Nachos, chili and beanie-weenies—it doesn't waste time, when there is no time to be misused.

(I love to waste time when I have nothing else to do.)

Above the microwave, a large Robert Abbott print of this guy hunting bobwhites with two liver-and-white pointers in south Texas. Next to the microwave, a little refrigerator, also atop the credenza. (Inside the credenza, field files:

Locust Ridge Field, waterflood papers, Larto Lake Field, North Blowhorn Creek Field, Giddings Field, and on and on.)

Newspaper blurbs: a picture of kids in Vicksburg singing Christmas carols in a rare snow, all bundled up outside the *Vicksburg Evening Post* building, thirty-five different "ohs," some singing loudly, some laughing, all the mouths open, all of them . . . no mumblers. A photograph of about a hundred watermelons, taken at some art gallery. Four golden and dried aspen leaves from a forest in northern Utah, October 1982. (Was it really four years ago? But it seems like forty.) They were ankle-deep that day, and like gold coins in the stand through which I shuffled, on a Wednesday three o'clock afternoon, alone, vacation. A two-by-three framed poster of a coyote looking straight at the photographer. A map of the Pecos Wilderness—Carson and Santa Fe national forests, New Mexico. A framed print of an American elk (326 of 750) by Texas artist Charles Beckendorf. Above that, a drawing of two red wolves (250/750) by same. A Polaroid of Elizabeth, taken in 1970 when she was twelve, sitting in the back of an ancient roomy car, grinning.

A postcard of five geese flying in some snow.

The hood emblem from my long-since-defunct VW Rabbit.

A picture of my youngest brother, B.J., trying to play a guitar when he was about four, and singing.

Coming around behind me: a huge bookshelf mounted to the wall; books, papers, stories, letters, folded maps. It groans. It is dark wood and handsome. A bottle of Mylanta, a brass armadillo paperweight. A spray can of horsefly bomb that I keep forgetting to take home to the farm.

Directly behind my desk, the drafting table, the battle-field. Currently unscrolled across it is a regional map of south Louisiana. Looking at Wilcox fields, and in between the fields. When there is a rush, an exciting emergency, I merely unroll whatever the rush is and work atop the regional map. And there may be two or three rushes, maybe eight or nine, in a week. Sometimes the maps are layered on my desk, like the rock formations beneath the ground.

Logs, colored pencils, erasers, straightedges, triangles, protractors, Magic Markers, calligraphy pens, little spools and ribbons of tape you can put around your maps to make them look spiffy and frilly: so much art, sprucing up what is only a science: the cold hard theories of where the oil is, backed up by the *feeling* inside you, and art: cheerful greens, lucid blues, fancy, tempting claret reds—lots of colors, and good cotton paper for the maps, which are folded and packaged in a leatherette binder! With executive gold trim! Please drill here!

The windows look out on downtown Jackson. (Rooster's Barbecue and mesquite burgers: thank God I can only see from here and not smell.) I have a friend at the Y whose building is across the street from the Holiday Inn's high-rise swimming pool. He's on about the twentieth floor, and can look out and see the secretaries from the Deposit Guaranty building sunning themselves over there on lunch breaks. And all the Mississippi-type pageants stay there, too—Miss America, Miss Mississippi, Miss Teen South, Miss Cotton. We're all insanely jealous of him.

More pictures; it is an art gallery. A rodeo star, beat, leaning against the chute after all is over, in the dust, old jeans and hat, new red-white-and-blue pearl-buttoned shirt. Old, old boots. He's thinking who knows what.

Photos of my own framed in rough wood frames. An abandoned truck up in a heavy stand of aspens, snowbound, rusting out, thirty years old, Colorado. Flynt sitting under a covered dock with his back to me in the morning fog, fishing on a still lake, a ring of water in the center of the lake where he has just cast his plug. An old house porch with a full moon coming up behind it through spooky-looking clouds. A board road winding through Lamar County, Alabama country, with the morning sun coming around the corner and running down the slick old planks through the woods like the Yellow Brick one. A little girl chasing chickens through hay in an old barn in upstate New York; the reception after a friend's wedding was at this farm out in the country. An eight-point buck, sneaking through boulders and cedars in Gillespie County, Texas.

Another bookshelf. A bottle of arrowhead fragments. The map racks: tubes and tubes of rolled-up maps, standing on end in various lengths and thicknesses, maps from all over the country, places my pencil has traveled. Oil beneath swamps, oil atop mountains, rivers of shale running through the woods. Atop the bookshelf, an ancient (1931) child's art book, very informative, actually. The text is a little basic, but the paintings are the same, they can't simplify those. *The Laughing Cavalier* by Frans Hals. (We can tell that this Dutch cavalier is quite proud of himself.) *Boy with Rabbit.* (Sir Henry Raeburn was the foremost Scottish artist of his day. He became both famous and wealthy.)

My turntable and receiver, logs stacked atop the big speakers. I do spend more time here than at home—I like to listen to music sometimes as I map. Albums on one shelf of the bookshelf, from slow to hard. A little blue bottle, found on a hiking trip with my grandfather. An old horseshoe, from same. Some acorns. Some rough rocks that I like, picked up in different places: not sleek polished fancy-dandy mineral specimens, but just good hard rocks with odd or appealing (to me) shapes and feel.

Another map rack. A forty-pound dumbbell in one corner by the door. A fly rod; the Pearl River bisects downtown Jackson only two blocks away, and in the spring it is a way to spend a shirtless lunch. A pair of snowshoes hanging on the wall.

A straw wreath lined with feathers: quail, pheasant, grouse, turkey, dove, duck. Mom made it for me. Whenever I went hunting and came back with some feathers I thought she could use, I would send them to her; it was fun to come home for each vacation and see how it was progressing. When I moved to Mississippi, she finally gave it to me.

It's a small office. But crowding it like this somehow makes it seem larger: I am somewhere other than the eleventh floor of an office building when I'm in here. It's a transitive trick: from the outer office, with its suits and ties and typewriters and water coolers, I go to the woods, snowshoeing somewhere or maybe paddling a lake, and then, breathing a little easier, from there I go down into the ground, however deep it takes.

It is a good place to work.

———————

Every geologist's office is different, and you can no more tell by just looking about how one of them finds oil or what his thoughts are about sand lenses and the grains of beaches and ancient mountains washed to the sea than you can by taking his temperature, by measuring his pulse.

And I would like to tell you that the best geologists have offices that are, well, busy—to argue that it is the signal of a busy mind. But the truth is that there have been giant oil and gas fields found by men in plain, white, sterile laboratory offices, and there have been giant oil and gas fields found by men in offices full of duck decoys and shotgun shells and aspen leaves and canoe paddles, and there have also been fields found by crazy men drilling for water, digging for gold, or shooting at deer. Just like in *The Beverly Hillbillies*. Really.

I can't name any fields like that, right offhand, but I know it's happened. I just know it has.

Look: you can't ever sell *me* on a prospect if you go and use a word like "scenario." To begin with, it's much too large a word, has nothing to do with either the simplicity or complexity of oil, or with finding it, but also, most damning is that it's an admission that whoever's using the word doesn't have stink for an idea of what's going on. The word implies a condition, an "if."

"Let's assume the following scenario"? Sheee-yit. Let's don't. Either tell the story the way you think it happened or don't. But don't try to con me. Don't try to take my money with probabilities, possibilities, and scenarios. Tell me where oil is. And why. Then if we're wrong, we lose. But I don't want to go into something questioning it, thinking: This is only a scenario. Basically it's obscene, in that it takes the beauty out of the act, the attempt. To find oil. It would be queer to call it a dance, but it is true, the drill bit does spiral, going down, spinning, around and around. Sinking through earth. Going backward. It's magic.

Someday I am going to drill my own wells. There is no geologist who does not dream of this. It is what you are after: doing it for yourself and not another. It will be fun. I will do it a little differently. Sometimes it takes awhile for the woods to close back in after you've drilled.

When we are through, Elizabeth and I will plant wild roses and lespedeza and fast-growing grasses where we have been. It will look prettier after than before. We'll put flowers along the board roads leading to producing wells. And bird feeders, wood duck houses. I'm not going to put up with any of this rusty cable and wire scraps and tin cans and mess lining the road. It's going to be like a park wherever I drill. It may cost seventy dollars extra, three hundred, a thousand and eighteen, but that is going to be how it will be done. It's the way it should be done, and it will be nice to see that.

_____

The phone company sent Elizabeth $252 yesterday, out of the blue. Her luck continues.

I love sitting on tractors and smelling their old oil smell. It sounds as if I'm drinking a beer but I'm not; it's really an orange soda, and the March sun going down is still warm on my left elbow. I'm not as far away from the interstate as I'd like to be, but it's all right. There's a nice breeze. It's the end of a workday and then some, and I'm still waiting on my little truck.

I'm on this graded hill, up above the repair shop, at the edge of a pine woods. Lordy, there are a lot of cars on the interstate. There's a fine haze between me and them; they move quickly, but their sound is only a soft hiss from here. Occasionally a motorcycle will blat louder than the rest. A crow caws behind me and to my left. Another one to his right joins him. People, secretaries and clerks for the repair shop, are walking down the hill to their cars. The men are still working on my truck. I can see them from up here. There are five of them, all circled around it, some down on all fours, moving around it clockwise. They look puzzled. My truck will not drive in a straight line. It is not the alignment; it is not a bent rim, nor is it a crooked frame. It is just an ornery truck. It has four brand-new expensive tires. No one knows why my truck will not drive in a straight line. I have stumped the experts. I continue to sit on the tractor and am vaguely amused.

Beyond the town, looking north, I can see the belt of trees along the Pearl River, which courses thickly through the town, and defines it. Cottonwoods flush, tinted in their tops, catching the flat and final part of the sun. The breeze goes on. There's a plane overhead.

I finish my orange soda. I know that at this point I should be compelled to make some statement about fuel oil, energy for farmers, heat for everyone, petrol for vehicles, power for industry, overhead lighting for operating tables in hospitals, and how airplanes are good.

I know too that the haze I see is partly due to spring pollen but also to the effects of people and industries, as well as myself, burning the energy that I discover. That there are things in the world that used not to be here, not so severely anyway, such as acid rain, acid water, and lead emissions in the snow on the North Pole. Oilcans on the bottom of the Pearl River, and worse.

There is only one thing you can do, though, and that is to do your job as well as you can, as well as it can be done.

I don't know what to say. I seem paralyzed by thought, or knowledge. Burning hydrocarbons causes good things and bad. The bad things are getting worse.

I know how to find oil.

I still want to tell you what it is like. I don't want to stop doing it.

Finding oil is sometimes like the feeling you get driving a little over the recommended speed limit on that sharp turn on the interstate outside Baton Rouge en route to Lafayette, when you come around that climbing corner pulling G's and truck blast a little too fast and then after that whip of a turn find yourself looking up at that space-age take-off ramp they call a bridge that spans, after the vertical climb, the Mississippi River. You've been driving along all morning on this pretty but bland, flat interstate, humming long into predictability, and all of a sudden there are all these surprise marks on the landscape: intense abruptions, challenges to the spirit. You find yourself almost racing up that bridge even before you've fully acknowledged its existence. You look down and see water. It makes the back of your hamstrings, and a wide zone across your chest, tingle. That is what finding oil is like.

Doodling around in the den of the villa at the yacht club (oh yeah, I'm a dandy. Tonight the North River Yacht Club and the kind of wine that tastes like water, and tomorrow night I'll be wallowing in the mud and mosquitoes of south Louisiana, logging a well in Terrebonne Parish) where I'm staying as a guest of Tom Watson, our fine attorney, before testifying at tomorrow's Oil and Gas Board hearing. Bill, our engineer, has been staying a few villas down, going over some of his own stuff with some guys from another company, concerning a unitization, I think. Anyway, he's been meeting with a bunch of guys, all of whom I've heard of or seen their names but none of whom I've met, and we're all going to go out to dinner tonight, coat and tie, at the yacht club, watch the moon on the water, etc. (And yes, I intend to order lobster, and roll it around in the butter, to boot. I'm a long way from home, and tomorrow night I'm going to be an even longer way, and I'm hungry.) Crickets are chirping; a rain earlier this afternoon. Cool air sits heavily atop the warmer air, making fog that lies along the ridges of the mountain.

Bill knocks on my door, come to pick me up. I let him in. He's red-eyed and kind of gasping.

Bill's an outdoorsman like myself. We hunt turkeys together, and have backpacked in Colorado on wonderful occasions. Being an engineer, Bill has to be cautioned, usually by me, against stodginess, but still he is an okay guy.

He seems agitated, almost frantic.

"Rick, before we go down there, I've got to warn you." He

is looking at me earnestly, his eyes wild, serious as a heart attack.

I look up with caution, with interest.

"It's bad down there, Rick, real bad. It's like an opium den. There's two guys from ——— Oil Company . . ."

I am making astonished sounds; I am agog. Bill, stodgy Bill, has floored even me. All geologists have boyish male-dog stiffness of attitude concerning other companies' geologists (Competitors! Surely not as able as our own staff!) but this is carrying it too far. An opium den! What kind of things does this mysterious ——— Oil Company do to find *their* oil? There is always the question, the suspicion, the comparing-your-company-to-theirs curiosity. You'll believe damn near anything.

I believed the two guys from ——— Oil Company were down there in their suits, seated on the couch, cutting smack on the coffee table as they talked about unitization. I mean, star athletes do it all the time, and these two guys were near tops in their particular company . . .

It turns out Bill was talking about their cigarette smoke. He's highly allergic to it and suffers terribly. For Bill, would that it *had* been opium; smoke is that bad for him. But he stopped at my horrified stare, and I explained to him what I was thinking, and we howled. We roared until tears came. And we went to dinner, and were still giggling slightly when we thought about it an hour later.

But still, you can never be sure what those other companies' geologists are up to: snooping around, prowling the same basins, the same state, trying to find your oil . . .

94

In the lounge at the ever-familiar car repair shop, waiting to get a ride to the office. A dozen other people are here in a similar predicament. One of those morning shows is on, too loud, furious interview, so loud it subdues everyone in the lounge except for two men sitting closest to me who are shouting back and forth about semiconductors. Posters and pictures of car undersides and tires and smiling mechanics looking up from opened car hoods all over the walls; empty Coke cans on the tables.

A lady crinkling her bag of peanuts from the vending machine. Something's wrong with my life. I seem to be a little tense these days! On the other side of happiness; shaled out, faulted down. Hmmm.

Core point is this: dramatic and precise.

Usually, as you drill down through the formations, your bit grinds them up into little flake-sized pieces, which are washed up to the surface by a circulating flow of heavy mud. The chips are lighter than the mud; they float in suspension. The mud circulates back up and is routed through hoses across a shaker that vibrates up and down (this after the mud and chips come through a contraption called the possum belly—damn engineers), and you can look at the little sample chips, scoop them up off the screen and bag them. But they're not the rock intact; they're just tiny pieces of it. You can sniff the chips for the odor of oil or gas, put them under a microscope and ultraviolet light, and make fairly accurate extrapolations from them as to what the rest of the formation might be like. This is what a mud logger does. He lives in a trailer next to the rig: lonely hours, a TV and radio and hot plate and many paperbacks. He looks at every batch of samples that come across the shaker, from start to finish, and does this rough analysis on them.

But when you really need the most accurate story you can obtain on a formation, top to bottom, with minimal disturbance, you cut a core through it rather than drilling it up. It is enjoyable, if you do not let it make you nervous. It's certainly a good and valuable thing to be adept at.

Coring is expensive and slow, which is why you don't always do it. Rather than having a bit on the end of the drill string, you have a long (thirty- or sixty-foot) core barrel, laced at the end with sharp diamonds around the edges. It's hollow like a sleeve and acts like a cookie cutter, except in this case the dough is rock and can be sixty feet thick. (If it's thicker, you cut two cores, or three, whatever it takes.) The drillers cut down through the formation, encircling it with the barrel, keeping it intact, sampling a long narrow plug (the core), and then, also slowly, they pull the drill pipe back out, stand by stand, until they get to the core barrel. They bring the barrel up then too, and knock the core out by banging on the side of the barrel with hammers. The core slides out like a skinny pole, wet and steaming, broken in places, but they piece it back together and wrap it quickly in foil to keep all the oils and gases and fluids trapped in it from escaping, evaporating, trying to rush out into that good light atmospheric pressure.

They place the core in numbered, long and skinny cardboard boxes that are rushed down the steps and into the back of a waiting covered pickup truck, engine idling, which then races off into the night, back to the laboratory, three, four, five hours away, no stopping. There it is analyzed immediately, no matter the hour.

It's the geologist's responsibility to coordinate this, make all the phone calls to get all the service people in the right places when they're needed. You've got to be good with phones to do it. Making sure the specialists who cut the core are there at a certain time; having the man who picks up the core be there at a certain time; having the lab analysts who look at it be at their office at a certain time . . .

But picking the point where you tell the driller to stop drilling, and come out with the bit, and fasten the core barrel on instead, and go back in with it—that in itself a tremendous expense of time and money—that's the magic. Drilling, drilling, drilling, watching, and then knowing, or thinking you know, that you are right on top of the formation you're interested in: a foot above it or less. Son, you had better be right.

And you usually have a lot of time to second-guess. You've been looking at drill rates and at the samples, but these characteristics change from well to well, even in the same area, and there can be faults, and the formations

will be thicker in some areas than in others . . . And all this while the drillers are coming out with the steady, fast, reliable drill bit, and preparing to screw on that slow, *expensive* diamond-studded core barrel, which is good for only one cutting. It will be ruined, whether you are right or not—$10,000 for that, plus another $10,000 in rig time. That's when you're standing there with your clipboard and notes and maybe a cup of coffee, and nothing else, and all the roughnecks know the pressure's on you, and they don't let up. They do absolutely everything in their capacity to make you second-guess yourself. For those twelve hours, they swear up and down that you're wrong, that they've worked on every well that's been drilled within fifty miles of here, and that none of the other geologists ever picked a core anywhere near where you have chosen.

Or they'll tell you you're already past it. That's the worst: going *through* your objective with the drill bit, milling it up, and not even being aware of it. It's not as if you can go back up the hole and *correct* your mistake; once you're through it, that's it. At least if you core too soon, and it's important enough that you get a core, you can go back in and try again. But neither of these practices is recommended. You have got to be right on the money.

Also, this is what they, the roughnecks and the driller, call you: Easy Money. The shouts go up the minute they see you driving up to the location: "Easy Money! Easy Money!"

It's like being the field-goal kicker. They don't need you, they don't need you, they drill for days and days, but then one day they need to know when to stop drilling, and you have to come in and be exactly right, down to the last inch. And you do it, and ignore their second-guesses, and when it turns out you were right, you can't even look relieved.

There was never any question.

Today, while driving, I discover that whenever I let up on the accelerator, the tape in my tape player skip-stutters.

It is April 21. Even at night, long after the deed is done, you can smell the grass that was cut today. The breeze is cool; I am drinking a Coke.

By now I hope that you are starting to realize that finding oil is no different from the job of selling insurance, or wiring electrical fixtures, or running a restaurant. And yet, somehow, these things—Cokes and honeysuckle smell— they matter more than you can imagine.

But I have lost it again—whatever I struck in me when I was thinking about telling you about oil, and then smelled the smells, and heard my tape player skip. The way I might have done if I were a farmer, or a baseball player, or an electrician. So that misses again, though maybe it barely breezed by you. Maybe you faintly heard or saw it.

If it wasn't such a magic and hopeful thing, I would be frustrated.

The sea. Maybe when you look at the sea, at night. Have you ever done that? That is the same type of thought as knowing about oil and wanting somehow to get it out of the ground. Have you ever looked at the sea at night, when you could almost hear better than you could see?

Things can happen to you anywhere, even when you're logging a well. When I get in the truck and start up the Trace to Vernon, to Fayette, I have no fear of boredom. I know that something will happen or that I will see something.

It is an old story: it happens over and over, wherever there are hills. A car or truck, older model usually, with manual transmission, parked without the emergency brake. Not a big hill, else the brake would be set. A hill just steep enough for the truck to roll down when the dog left inside it (man inside shopping for groceries; woman in the courthouse paying traffic ticket) grows hot or restless and hops around on the front seat. Knocks the shift arm into neutral. Away she goes. The dog puts his paws on the steering wheel to catch his balance. The wind begins to rush through the window as the car gains speed. This feels good; the dog's mouth opens, he pants; it looks like a grinning dog is driving the truck. I got hit by one of these things in Fayette, Alabama, once. Again, I was on that umbilical cord to the office, seated in my own truck downtown, calling from the pay phone (again, the only one in town that *worked*), and at that time I still had a rearview mirror. I was lost; I was asking directions to the well. I saw the dog and truck coming toward me. I dropped the phone and started the truck and popped the clutch, lurched forward—not in time, of course, though I had made it about ten or twenty feet when the truck caught up with my own and sent me skidding farther along. A game of bumper cars, a loud noise on a busy noon Wednesday in Fayette. For a long time after that, investors, secretaries, even Mr. H. himself would get very excited when I'd call in with even the most routine information, picturing in their minds at that very instant another disaster hurtling down the hill toward me as I spoke: they would interrupt my conversation to warn me to keep a close eye out. And there was another dent in the truck to explain. I never fibbed; I never made things up. If they asked how the taillights got knocked out, how the *back* bumper got bent that way, I would have to tell them, sadly, I was hit by a dog.

Another time I was in Pachuta, Mississippi. It's south of Meridian, in a basin entirely different from the Black Warrior—what they call the Interior Salt Basin—an embayment, Jurassic in age (Smackover, twelve thousand to fifteen thousand feet deep), where the Gulf of Mexico (even then!) once came way ashore. Structures and faults and porosity in the Smackover lime, sometimes sand, old beaches, oil. It's where Mr. H. made his first big big big money: he found one field, assumed all the others in the area would act the same way, and ran in a forty-mile east-west string of successes that proved him right. Little mirrors, field after field, all along the same fault system, all the structures holding oil, all the structures huddled right up against the fault. Dozens of long, skinny eggs. Millions and millions of dollars. It made him fond of the area for a very long time.

We drilled several wells there, in that part of the basin,

searching for another fault system like that one. We never found one, although we drilled a lot of deep dry holes searching, sometimes fooled or goaded by traces and stains and tastes and odors of oil, old oil that had turned to asphalt or was present in only that one well—and eventually we drifted out of the area—investor (as well as our own, less easily admitted) discontent: we started looking for runs of faults in other areas. But back then, we were drilling there a lot.

There was an Indian, Choctaw I think, in this little town of Pachuta that had all these wells around in it, back in the woods. He may have had an oil well, I don't know. He didn't have a car. He drove a wagon pulled by mules. He came from far back in the woods, once a week, to shop for groceries. The store had a wooden floor. He wore a rounded canvas hat with a turkey feather in it; the wagon creaked and was about a hundred years old.

I was picking core point. We were making a bit trip, and I had a lot of hours to kill. I left the rig, drove into town, and found myself sitting outside on the steps of the grocery store waiting for the Indian. I talked to him; he let me take his picture and ride with him all over town, running his errands. We drove through the country, and the mud logger drove up behind us on one of the roads. I turned around, recognized him, and waved.

There was a lot to answer for when I returned to the rig. Yes, I had been riding around with The Indian. ("He's crazy! Did you know that? That Indian's crazy!") The geologist who was picking the million-dollar core point was off riding in a wagon with a crazy Indian. I hadn't *told* anyone, but the fact that it had been *reported* made it unquestionably odder in the Mississippi sense, the small-town way.

Once, in Pachuta, I ripped the only phone in town off the hook. At the time, there was a serious-rank, top-secret, ultraimportant wildcat: a new fault block, and I had to call the office about it. It was raining, so I pulled up to the phone and dialed from inside my car. It wasn't the kind of phone booth where you were supposed to do that but I did it anyway. It was right outside the city hall (a little wooden room of a building, like a child's clubhouse; it was also the post office from nine to twelve), and there were some engineers and lawyers from Another Company hanging around that phone like vultures. We had just logged and had about four feet of marginal pay, and they had lease positions in the area too, but there was also some open acreage available, and if there was pay they were going to race right out and buy it up before we could, just on the news. I knew they were going to listen to what I had to say; that was why they were standing there.

The whole time I was talking to our office, I was watching them watch me. It was nervous-making. I was trying to concentrate on two things at once, and sometimes I am not good at that. When I finished I rolled the window up quickly—a light rain had been misting my face through the entire conversation—and drove off, forgetting to hang up the receiver. Or maybe I was just pulling up so I could open the door—I think the booth was blocking it—so I could get out and *then* hang up the phone. At any rate, when I drove off it rooted the receiver and cord out of the phone like a carrot being pulled from loose soil. I heard all this noise, looked down, said "Oh, no," and just kept on driving. I think they took that to mean it was a very good well. They didn't know *what* to do. And it wasn't as if they could call their boss to find out, either.

It is raining hard. The stereo is playing. I am alone. All the windows are shut, five o'clock in the evening. The rain is thundering, coming down hard. The stereo is up loud. I'm completely happy. It feels too easy: like walking in a dream. Surely I am missing something. It cannot be this easy. Happiness is supposed to be sought after, complex, to be found only with the greatest amount of cunning.

Water roars off the roof, and I am dry.

Later tonight I will fix coffee.

This is a thing oilmen (masc. and fem.) do: they put oil in sample jars, little solid glass cups with screw-on caps. You can unscrew the cap and smell it. You can set the jar on your desk. If it is a high-grade condensate, it will look like Orange Nehi and smell like unleaded. If it is low-grade, twenty-degree asphaltic, it will smell like tar and be sludge-black. Of course, if there is H_2S in it there will be a sulfur smell, like rotten eggs. This sour gas is very lethal, even at modest concentrations.

It is great to hold the oil that you have found in the field in your hand. The sample jars are the same ones used for bottling the sidewall cores before sending them to the lab for analysis (the screw-on lid prevents any gases from seeping out). The bottles are frosted on one side. You can write your sentimental, dizzy ideas on this frosted surface with a fancy pen, though just as many do it in the field, at the first well test, with a nubby pencil. The name of the field (or wildcat), state, county, date, formation, and depth. I have a bottle from Atascosa County, Texas: a wildcat, the first well my father ever drilled by himself, fifty-three hundred feet down. It smells good. It's blackish green, and I've had it almost twenty years now.

I don't know if I'm really communicating to you the strength and importance of these little bottles when they have oil in them.

You can have a medal from the Olympics; suppose the fifteen hundred meters was your event. Or you can have a photo of yourself hitting a home run in the World Series. But you can't hold those things and you can't put them in a bottle and see them after they are gone. Immortal as those other things may sound, capturing energy is really the most magnificent experience. The bottle of oil really does seem alive, empowered. Bigger than you in its two-ounce jar.

Hold it up to the light. Tilt it slightly. Lower it and unscrew the cap. Smell the oil. Touch it with your finger.

When no one else is in the office, put the cap back on and hold the bottle up to your ear. Picture an ancient seashore. A world so different from the one we are in now it is frightening.

No more Cokes! They are outlawing them! Not enough *sugar!* cry the company's researchers. There is a tight rein of panic around our necks, and Elizabeth and I are driving through the country, stopping at all the little stores that haven't sold out yet and buying Cokes out of the refrigerators, the iceboxes, taking crates of them from the storage rooms. Hot summer skies overhead. The crates clunk and rattle as we place them in the back of the truck. We drive on, looking for a new country store. There's a secret we've discovered: anything with a D stamped on the cap is an Old Coke. We are wily, discerning, seeming crazed to the storekeepers who possibly have not heard the news.

"Yes, we'll take that crate, and that one," we say as we examine the tops of the bottles. "No, we don't need that one, thanks . . . Yes, that one over there will do. That will be very nice . . ."

We have grown accustomed to sitting on the back porch and drinking Cokes, watching the sun set, watching the condensation bead on the bottle, then break and roll down . . .

We have a mission. By week's end, we have over eight hundred of them, and are feeling a little more relieved. We don't want to say "I can't" to having Cokes on the back porch in the summer.

Cokes and grizzlies: it is going to be that kind of a summer. In late June and early July I will go up to Montana (and Idaho) to write about grizzlies, of which there are only 1,200 left in the United States. Now, in May, there are 1,375 Coca-Colas (old formula) stacked in my den. There is only one Elizabeth, and she is coming over for dinner tonight. (Salad. Cool air through the windows.)

I turn the lights off when I leave a room. You've got to respect something you know someday is going to beat you— like youth—even though at present you may seem to be in control of it.

The earth is strong. You learn this. Not that you ever suspected it wasn't. You just get the opportunity to feel blown around and swayed more, working with it.

Suppose you have a gas reservoir or, say, one with gas and water, the gas floating and bobbing up and down above the water. You perforate a little too far down in your well bore, and one day the water can't resist: the gas above it is flowing to the surface, being rushed out into the pipeline, sold to companies and shot out to the nation, pipelined underground still but not nearly so deep. And the water too, with the gas, seeks the lower pressure, the easier times, the well bore. Upward it comes, to blue sky, fresh air, green growing things again, and sissy-pants baby-piddling nothing atmospheric pressure. You might say it likes it.

It's sad to say, but once this happens ("water encroachment," "gone to water," "watered out") you're rarely strong enough or lucky enough to change it. You can try to fill with cement the perforations that opened the reservoir, and then start over, reperforating, only not shooting quite so far down this time. But the earth shrugs at a sack of cement, at two sacks, at three, and the water remembers. Paths taken by the earth are not easily reversed by anything, and certainly not by man. It is hard to change the paths, really, of even the slightest of natural things: a relationship, a moth to a light, a dragonfly trying to get to a pond, a dog that chases flies. How are you going to tell an old ocean that has broken through a gas cap to stop remembering that blue sky, and go back down?

The Hubbard Butane truck came; it was the first time I'd ever seen that operation. They unrolled a hose, screwed it into my tank, and flipped a lever on the truck. There was a number-clicker thing that showed how much was being dispensed. I got a hundred gallons. It was pretty basic, not much suspense or anything to it, but I liked watching it, and now my tank is full.

Geologists, if they work a new basin long enough, and if the basin holds enough oil and gas, get to see the basin "mature." It's not the basin that's maturing, of course; it's the same as it was a mere hundred years ago, or thirty, or eight, or however long it has taken for the geologists to decide it is mature. What they mean is that there is now enough information on it, enough control—seismic lines shot, wells drilled—for them to have a very good handle on it. They mean the basin is in the peak of its years, exploration- and development-wise. They mean it's at that vague point where you're more likely, during the history of drilling in that basin, to complete a well than you ever could before (or will be able to again).

When a basin is "young," there isn't enough information. Even in a prolific basin, such as the Black Warrior, the success rate might only be around 25 percent. Paradoxically, up to a certain point, the more oil and gas fields found, the easier it becomes to find others. You would think that every time one was discovered, it would make it that much harder to find another—that there'd be one less around. But before you reach a saturation point, it doesn't work that way at all. Because that's the single best way to discover oil: to figure out how it is trapped, under what conditions, nearby, and then look for another area that has similar conditions. To the north of the nearshore bar that is Fairview Field is evidence of another nearshore bar. This is drilled, and becomes North Fairview Field: A little northwest of there is a sprinkling of dry holes that could be the edges and sides of *another*, similar nearshore bar, about that same size, length, and width, running in about the same direction, parallel to the old shoreline. This is drilled, and a new field is discovered. There are a few golden years in the maturation of any successful basin where, truly, if the homework has been done and all the facts studied and learned, it is . . . easier.

After a while, of course, the saturation point *is* reached; it gets harder to find new fields. Because finally a large percentage of the fields *have* been found. It is so like middle age that it is depressing. Basically, one day you just notice— though it may have been going on for quite some time before you admit it—that you are not finding oil wells as frequently, or with the success rate that you once were. It's not an immediate thing, but it's part of the phase too, part of the cycle. "Overmature" is the term for it.

There is always hope of rejuvenation, however; hope for a comeback, once the glory days are gone. There are basins and basins, many many many all over the country, that used to be hot in the twenties, the thirties, the fifties, and were drilled up, and now, years and years later, a new zone is found. Sometimes the zone is deeper, though occasionally

one is "up the hole": a zone that has been there all along but that no one knew held oil, held gas. Sometimes, as in the North Texas Basin, the zone exists just a few hundred feet below the ground! A basin is always capable of making a comeback, as is a geologist down on his luck. As is everyone. All it takes is the usual factors: work, research, another look, luck, the desire for the comeback to take place . . . whether spiritual, in the hearts of men, or economical, in the oil industry. There must always be a motivation for such an event to take place.

Oil is an American industry. Not to say that no one else has oil, because they do, and more of it than we do, for reasons of sedimentary whim. But when people were burning coal in the 1800s, we were finding oil in Pennsylvania, Texas, and Oklahoma. We found it coming out of the ground, and it got in our water when we were drilling shallow wells by hand. Rather than finding it a nuisance, we turned it into an advantage and ran with it. We have taught the rest of the world the best way to look for it, the best way to find it, and the best way to use it. It's cleaner than coal, more powerful and economical.

Oil is maturing as an energy source. Someday it will be too old or too scarce, but right now it is the best, and, as is always the case, it would be wise to appreciate and take advantage of its maturity, however brief, however extended. But to also admit, when the time eventually does come, that it is overmaturing.

But what a loss to turn the oil in early! To give up too soon! To preclude rejuvenation. As long as there is motivation, however, this probably will not happen.

To overtax the resource: this confuses me, this windfall business. But perhaps there's something I'm just not seeing; I probably just don't have all the facts. Maybe my philosophy's all wrong.

In the northern part of the Black Warrior Basin—what is called updip, the final striving reach of the oil and gas, always trying to climb—in this part, a dry hole tells you where the oil *is*, in addition to where it isn't. Up here where the oil sands are snaky, fat, and prolific, as numerous as catfish, a dry hole helps pin down the outline as well as a producer does. Sometimes it helps even more, because if you are square on the edge of a sand, the log is going to reach a certain way, and if your eye is sharp and you have seen that certain reaction before, you might recognize it.

———

Looking at the map every day, every week, over the years of your life, as more and more wells are drilled, is like watching one of those Polaroid pictures developing. The oil down there, the sand bodies, the faults, never change, never shift position, certainly not in our winks of lifetime, but the picture is ever-shifting. It makes us look more than a little foolish and indecisive, but that is the best we can do. The only strong thing to do is keep drilling anyway, reacting, adjusting, changing, trying to catch up to the geology's silent superiority, its mute, perhaps bemused tolerance.

Just because there is a chance something may not happen is rarely a good enough excuse not to try it. It makes no sense. It throws a wrench in the tension of life's processes that string momentum along on a beaded chain. It's like walking off a football field in the first quarter because it's your football, or because the other team has scored a touchdown. It makes no sense at all, and for all practical purposes, stops things. You still may be living, and breathing, and eating, but you've halted something, and it is very important instead to keep this thing going in you for as long as you can.

Men sitting across from each other at a conference, leaning forward, palms on the table, looking each other in the eye. One is speaking softly, earnestly. You can feel the sprawl of earth and depth of sincerity trembling underneath his palms. He wants this man to believe as he believes. His words are so measured and loaded and forcefully gentle, so gentle. His eyes do not break contact.

". . . The water has been coning up, leaving several desirable locations undrilled on all flanks of the structure . . ."

I see all of this as I pass by the open door with a Diet Coke. It is 2:30 P.M. and is raining outside. I'd buy into the deal based on the way the guy's palms are placed on the table.

But you don't see that too often. Not too often at all.

There are people I know who dabble, who want to write—no, who want to be writers. But they're married, or have children, or have a job, or watch the news. There's no time. Or they need to be inspired. They wait for it, and it comes about once every three years, and half the time in those instances they're without a pen, or think they'll remember it—their inspiration.

They think I yawn a lot and sit around on my fat butt in the country, spinning yarns and poems and occasionally planting a garden or two, sipping a beer, when actually I'm caught way up in my writing.

Well. Maybe they're right.

———————

We lay in the hammock together today. Her yellow shirt was the color of a banana. I didn't go to work until ten. I was burned out. I didn't lie, didn't call in sick. Just told them I'd be in around ten.

You can't find oil if you're not honest. I'm not sure how to explain this.

I'm not saying the way to find oil is to stay in a hammock till ten in the morning, either. I'm just telling you that that's what I had to do this morning.

I found two dogs, tiny puppies, twin hounds, hiding in the weeds beside Old Port Gipson Road. I couldn't tell you the series of left and right turns I took to get there; it was a random, whimsical route I'd never taken before.

There was a third one, dead, in the center of the road, that had been hit by a car. The others were in the weeds looking out at it. They jumped back as I drove past: the sound my tires made on the wet pavement, a two-lane road through deep woods near the river, and I was disturbed, seeing them like that. Floppy ears.

I looked in the rearview mirror.

The two puppies had jumped out of the weeds and were running side by side, as hard as they could, down the center of the road after me. It was as if they were positive I was the one who had abandoned them. It might have been they were dumped by someone with a truck like mine.

I didn't have to go far to decide to turn around. I went through a curve, made a U-turn in the wet, empty road, and when I drove back around the curve, in the other direction, there they were, still coming: tongues hanging out, running for all they were worth.

They had worms and mange—probably why they'd been abandoned—and ticks and fleas and burrs. Their bodies were skeletal. One rolled over when I bent down to pick her up. She began peeing wildly, a mad golden fountain. Peeing Ann, I named her instantly. The other one darted away, then came running back, barking, trying to protect her sister. She ran right up to me and barked savagely but then tucked her tail and ran for the woods when I bent down to scoop her up. I had never seen a puppy being protective before.

I carried the still-peeing Ann over to the truck, waited for her to stop, and put her on the floorboard. She promptly flipped over on her back, like an opossum, and put all four paws in the air.

I went into the woods to find the other dog. I hunted her for about thirty minutes, right until dark, before giving up. If she was that wild, she probably would be better off in the woods anyway, I rationalized.

There was less than five minutes of light left. Coming back out of the woods, I passed an old abandoned house, the kind with broken, dust-opaqued windows, caved-in roof, and sunken, splintered porches. The kind that is vine-covered and has trees growing up through it. The kind that, as a kid, you'd be scared to go into.

As I said, it was that gloomy hour right at dusk.

I went in. You never know what you will find in those old houses. I was nervous.

I began to find dog poop here and there. Rotting couches and a TV with the screen bashed in. Closets that I was frightened to open. A dank chimney, an ominous kitchen invaded by spiders and their large webs.

She was hiding in the last dark room I looked in, in the far corner, burrowed under some newspaper. I scooped her up and hurried out.

She was so small that she could fit in a shoe box. She tried to crawl under the accelerator pedal.

I was closer to Elizabeth's than to my place. I went by

there, and we gave them milk. They paused before the bowl, then leapt upon it like hockey players in a face-off. The milk was gone instantly, as if vaporized. We gave them some more.

I named the wilder, shyer one Homer, despite her gender, after Homer Wells, the orphan in *The Cider House Rules*. I'd just finished reading it, and was impressed with it.

On the way home, they slept on the floorboard, with their arms and legs tangled around each other as if for warmth, Homer's head resting across Ann's back. Milk on their whiskers, round stomachs. It is hard to describe the feelings they brought out in me, and they are just dogs.

At this stage of the relationship, I could write about the dogs constantly, but a tape recorder and mike would be more suitable. A very muddy well to drive to last night: a three-quarter-mile charge up yesterday's clay out of a flooding creek to the top of the hill. Driving all last night with the dogs bundled up in an old blanket, the wind blowing so that more often than not they were covered entirely, blanket flapping. Dogs in a blanket. Stowaways.

(Lucien, a character in Tom McGuane's *Something to Be Desired*, sneaking into a hotel with his spaniel: "C'mon, Missy, you can do it, the hunchback under the overcoat trick?")

The startling sureness, this idea of picking them up when I saw them by the road—I have been seeing dogs along the road all my life—yes, they are for me, maybe more so than any other dogs ever will be. (Dogs consume me, literally. I've been bitten in the leg and arms by a Doberman, on the butt by a boxer, on assorted fingers and toes by basset hounds, and on the wrist by a pointer.) But to draw the reins in and examine why this good thing happened to me, so that it can be done again, more often—in tennis, in choosing drinks at a bar, in buying a car, in taking a fork in the road. I did the dead-solid absolute best thing for me I possibly could have done, and I didn't mull it over, consider, or speculate—I just reacted. What sequence of events had come before that? What had my frame of mind been? What had I eaten for supper?

Swallows or swifts, I don't know which, are building a nest on my back porch, daubing mud into a corner with their beaks. Eggs, then baby birds will arrive. My farm is becoming a home.

I am eating biscuits and milk gravy, having just finished the eggs (eight), hash browns, sausage, and fine coffee, in Guin, Alabama. I am looking at the leftover bowl of gravy and thinking how much the puppies would love it. A week ago I would have been mooning about how Elizabeth should have been here. Well, she still should. And I am only being practical, I tell myself. She doesn't really like gravy.

———————

Finishing the coffee. Emmylou Harris is playing the guitar slowly on the radio, and singing. I've never been happier than I am this morning. No reason why, and I can't define it. The serendipity of finding those dogs bothers me a little: not at all unlike drilling a well and finding oil or gas in a sand where you hadn't expected it. As I approach twenty-eight, I'm learning that you can't map happiness. You can only recognize it. And run pipe on it, as it were; produce it.

Beat, exhausted, owning nothing of this Guin well, I am putting in hard overtime. And I'm still able to adjust my stance and be happy.

All it takes is a weak radio station and breakfast and a ceiling fan and kitchen noises and puppies asleep in the back of the truck and Elizabeth and me being healthy and a white T-shirt that fits loosely, and mud drying on my boots. Yeah, go ahead and map that one if you can.

The light on my cotton sheets in the sun room looks like the light in an old movie from the thirties; the fifty years in between might not have happened. I go into the kitchen to get a glass of water. Very few people know I am here, and no one knows what I am doing: holing up. There is a breeze, as always, and a bird is outside feeding her crying babies in the nest. The landlord's wind chimes are tinkling.

I'm becoming a better fit. I can feel time working on me, sanding me, and if I do not struggle, I will sift into place, become part of the formation. (Is this what I want? If I desire another formation I had better hurry on.) Today, at least, it seems to be a noble cause. Who knows what our formation will someday hold, trap, and claim? I hope it will be valuable.

Here is a summer sentence from a brochure resting on my cedar coffee table (ten dollars at a garage sale; a beauty, just right):

Ramsey Springs. Take out at Highway 57.

Some days, all words look like poetry. The wind chimes bang together. Certainly it must be a goal to acquire a smoothness of fit, an aging.

Cheese. Wine.

P oems. Women. Dogs. Bookshelves. Canoes.

I want to slowly, eventually fit everything.

I'm working on it. I used to get angry so easily. Now it just makes my stomach hurt.

Meadows, pastures, fields, and prairies: nothing can hide. The wind moves across them like a balm.

Sunday. There is a butterfly in my kitchen. He is on the inside of the screen door. I pause to examine his colors: sultry eye-shadow purple, ovals of flaming mist, a few white pearl pinpoints, a hearty mustard thinning of junglelike stripes—then quickly I let him out the door. He springs into the air, rockets off into the pasture. My bookshelves groan; it makes me very happy when I touch some of the better books, much less read them. Sometimes I cry or punch the air when I get stuck in my writing. Falling in love can be a cure sometimes, but it can be a shot in the arm only if you don't *need* it. Nature's good old paradox: Drill here.

Driving: rabbit bounds along center of road and I slow down so he can escape his panic and remember to get off the straight and narrow road and back into the woods.

Mobile-home skeleton back in some trees, melted like a candle. I'd never live in one. Better to do like the Wyoming roughnecks during the oil boom in the early eighties, when money was thick and no housing was available anyway. They'd run a mile-long extension cord out into the prairie, buy a dining room suite and bed and appliances, and set up under the sky. Never rained anyway, out there.

128

Corn. Houses with porches. Mailboxes shining silver in the weeds along the road. Happiness: it is a thing to be lassoed, wrestled. No milksops or lightweights allowed.

Lacing shoulder pads in high school before a game: standing across from your partner, your buddy, lacing them up tightly and gingerly, getting them right, as if pinning on a boutonniere, and then backing off and slamming them as hard as you can, bringing the bottom of both forearms down on top of them, several times, ritualistically, to be sure they are secure. So that your partner will be protected (as well as in a proper, aggressive frame of mind for the game). Hitting him to protect him from injury; it was so very simple. I will try not to panic, the next time I am just a little sad.

Sometimes when I go to a catfish place by myself, just feel like getting out of the house after several days of silence and writing, I need to remember how truly good it feels to be totally alone (dogs asleep in the truck, waiting for me). All in all, I'd rather she was here much more than not, but there can and will be good times without her, too. That comes from within and must be preserved, I think, or it will disappear. It is simply not in her to commit.

It does no good to think about it, and worry. It is a simple deal: enjoy the times with her, and then when Europe, or tiredness, or a ten-year cruise to Nepal comes, remember how it feels, some nights after hard, isolated work, to call the dogs and get in the truck and go into the nearest little town for catfish. I need to keep that salted away: I know that someday I will need it, and it will be better for me to have it than not.

In the meantime, the horses graze, the dogs fight, we drive to Tallulah for daiquiris with the windows down, things remain the same, changing only so slowly that we do not see them, do not call them "changes" at all, no more than we call a beach "sandstone" or a jellyfish "oil."

Knowing what I know and accepting it, her nature, having learned it, is not the same as having another well to move on to, a new prospect, when you finish with the one you're drilling. It's not like knowing for sure there'll be pay, or for sure a sand will be there. In this case it is not like geology at all, and it's not like growing old at a faster pace than the earth, either. In this case it's not like anything, and I am flying solo, making up my own rules, going into a

place that is only mine and hers, for there are no books on
Rick and Elizabeth, no past histories to repeat themselves,
and remembering to head to town on my own every now and
then is one of those rules.

It would be easy to be pious. We have done it with peace, let it slip away from us again and again, and with forests, birds, fish, and mammals. We have our reading glasses on or, better yet, are looking at a mirror, not paying attention to the signs. Things are dropping back, falling away; we will suddenly (or so it will seem) be by ourselves. Nearly always, everything deserves the thing it earns, good or bad, in the category of fate. We are working hard for our disaster, it seems.

There'll be suffering and the rest. Disease. I am being callous.

The present glut in oil and gas is about four weeks long. It never gets overbalanced by more than that. Yes, prices are low. There is a surplus.

A surplus in the puny tanks. What about beneath the ground, in the earth, in our puny abilities to reach it?

When they did away with the Old Cokes, the day they stopped producing them, there were plenty of Cokes on the shelf, were there not? Perhaps even a glut, to someone interested in buying a lot of them? Hell, I bought a thousand, no sweat, and at a good price.

The fat and easy areas have been discovered. We, the world —not just the United States but the petroleum strongholds such as Mexico, the USSR, and the Middle East—are pulling out of the ground twice the amount we are finding. Each year.

We know how much we are finding. We know how much we are using. In recent times, the best year was back in the fifties, the worst was last year, and the second worst, the year before that. The third worst, the year before the year before. And on and on.

Using more and finding less. It shouldn't be confused with availability in the tanks, the greed end of it—What Can I Get Today?—because that is the politics of nations and their immediate and day-to-day need for cash flow; these days, that is *all* that governs how much oil is available and at what price.

Go beyond that, under the greed and dollars of it and into the purity. How many traps of ancient reserves are left, and how long will it take us to use, at our known rate, our known requirements, this projectable quantity? You hit zero, every well in the world a dry hole, in about sixty-five years. Do not think it will be a pretty sight.

I'll be ninety-two years old, anyway, and quite possibly wandering the woods with a sharpshooter, a few lengths of pipe, a truck full of empty, hopeful drums. Horses or mules will pull the truck. I will have dentures and will be remembering old oil fields I have found.

It's not like gold, silver, trees. It disappears. It is here only once.

It is amazing to wander through oil fields after a wildcat discovery and see all the energy, all the bustle, trucks driving in and out, new wells drilling, flares burning, bright clean tanks shining. Knowing that within your own lifetime

—not your children's or children's children's, but yours—
you will see a picture more different than could ever be
imagined . . .

Flocks of them darkened the sky—ducks everywhere.
Passenger pigeons went from billions to zero within ten
years . . .

To drive on through those busy oil fields, as in a time
machine, where the doomed present does not know its
future. To go to a movie, cut the grass, go to sleep with
the window open and a breeze blowing in. What you will
probably someday look back on with much hunger as a time
without troubles . . .

The world is so thirsty for oil, uses so, so much. We are
down to the last thousand Cokes and there is no telling
when we will recognize how few we really do have left, or
how much time there will be left to do something about it.

Y̶ou can't be wishy-washy. You can't *not* believe that every move you're making is the best possible one.

The kind of geologist I like: his face will cloud over at the hint that he "should have done it differently." The thought must be dismissed even before it is formed. He or she will be the first, absolute first to tell you he was wrong when this is so. But you can never make a great geologist believe he should have done it another way.

Occasionally, you will meet a geologist with an ego, who acts as if every lift of his eyebrows should be chronicled in a little black book. A *leader*—in a profession of men and women who cannot be led, and will not. They are in the wrong business, these salesmen of self. Do not suffer them. Ring their doorbells at night and then run. Cast them out. They make the earth and the oil that is down there in pockets and cracked fissures less warm. I like to think that when you are entering the freeway and your car hesitates, or when you shut your engine off and it clatters unexpectedly, this happens because the fuel harbors a few drops of oil found by one of the ego-geologists.

There is a right way to find oil, as well as a wrong way.

This is one of the things my father has given me: a 7½ percent working interest in a good little shallow gas well out in Zavala County, Texas. My monthly income check runs about thirteen hundred dollars; my expenses (a 10¼ percent billing interest) go about four or five hundred a month.

Naturally, I would not cry too much if the price of natural gas went up. But forget this talk of money. In the autumn at my rented farmhouse in the country I can hear owls at dusk. There is a yellow porch light. I would like to be a fat boy and live on this eight or nine hundred dollars a month (travel across Europe, camp in Montana, write, drink beer, strum on the guitar, retire in good conscience), but I couldn't.

Perhaps when all of the Black Warrior wells hit, and all the wells that exist in my mind do too, I can do some of that stuff. But I doubt it.

Very few people quit anything entirely. There are diets and divorces and ex-flames and favorite hobbies, but to *quit?*

A sound I've never heard before on the farm: a motorcycle, a long way off, going around curves and up and down hills, on Dry Grove Road, two miles through the woods.

Spring: Chuck-will's-widows.

Summer: Cicadas and doves, up until dark. Crickets, and silence after that.

Fall: Owls!

Winter: Far-off sounds! Cars on roads, and truckers . . . no leaves on the trees . . . other lives . . .

136

Well duty can be brutal, but coming back to the office in the morning after six days of lower-back driving and dust and no showers, coming in wearing jeans and an old shirt and a six-day beard in front of all these delicate, exacting people (secretaries, accountants, even the other geologists)—despite the tiredness from being on the road, this is worse. The tough part *really* begins now; all that exhausting stuff was just the warm-up. Now you are shoved into a supercharger and rolled out, faster and not especially in control.

You're bone-weary. You've been locked into the rhythm of enduring, not faltering, not sleeping so you don't miss anything, and then when you get back to the office they hand you a baton and ask you to accelerate (we need a little sprint action now, marathoner). An entirely different mental framework. You learn to do it, but it is hard on your tendons and ligaments; the ones in your mind as well as in your body.

And they go home at five and shower. Sit on the sofa. They have their fifteen and a half hours till next morning free, and they take it for granted. That is the worst: being jealous of that. Because you're learning things they're not, seeing things too. It's just that the jealousy and resentment try to come in anyway, like wind through cracks in a cabin.

The worst is that they ignore your doing this marathon stuff. If you do beat the jealousy, and not show it, then they think it's not there, or else that it's not a struggle, and therefore not that hard. (The marathoning, even the sprinting, is somehow different for *you*, because you're strong enough, or "used to it.")

I am whining.

But it's a bitch. If your boss is in a foul mood for one reason or another, and you've just come in bleary-eyed after having done a week of well duty, and he snaps at you, it makes for a helluva Job Conflict.

———————

This is a sweet smell: sorghum and corn and sticky molasses oats in the heavy brown feed sack. It sits in the back room of my house so the horses won't break into the tack room anymore. It makes my entire day smell good.

Some summers it seems as if it will never rain. A letter from a friend in the mailbox; purple thunderheads that build up in the heat of the day, even generating a little wind, but they always blow away. Tea with sugar in it at night. To break the spell, a night out on the town. Crystal Springs, population 1,000. Fried catfish at Sudie's: long picnic tables, church style, red-and-white plastic tablecloths, bare light bulbs hanging from the rafters, so hot there is nothing to do but eat food that is even hotter, so that when you stop, you seem to cool off.

You walk outside to the huge parking lot beneath the stars. People come from all over to eat at Sudie's. It is out in the middle of the woods, even outside the Crystal Springs city limits. The smart ones drive thirty miles from Jackson to do it, in the summer, on a Wednesday. Do anything to make it rain: go out of your way, break your own routine, to convince the weather to do the same. Stop the no-hitter. Make rain fall.

Who is Sudie? Is she the owner? The cook? What a lovely name.

My best friend Kirby and myself, college students at a football game. It was an exciting game, but of course we weren't interested; we were looking about, pointing out odd and unusual people, a whole stadiumful to choose from, a wealth of different lives.

"See that man with the beanie cap?" Kirby asked, looking across the field.

139

"Where?"

Kirby pointed. "The one standing up."

Even as he spoke, seventy thousand people leapt to their feet: an exciting play down on the field. We looked at each other and howled.

Good things can happen to you in any instant. Doubtless there is a scent that makes good things like you, come to search you out. I think it probably has to do with this fitting concept: the settling, sliding in. Not being off-balance, but defining yourself, rough edges and all, and then securing, locking, battening yourself down, waiting for what will happen as events try to round off your rough edges.

I like to dig postholes. You can feel sun on your shoulders, can use your muscles, can practice a technique, and you can be under the sky and see for miles and miles.

Then, too, there is the fit. You lower the poles into the holes. You chink and wedge sod and stone around the edges, and tamp. The post is solid in the earth. Something you have done has become part of the earth. There is a spirit of cohabitation because the earth does not (seem to) reject your fence post. And then you carefully, again in this same spirit of friendliness, string the barbed wire, tightly, like a promise.

There. You have contained the prairie, the field, the meadow. The wind blows through it. Weeds grow on either side of the fence and under it, as if nothing were there. The prairie lets you keep your horses on it. There is very little you can do to the prairie to make it not be a fit. It is very hard to change a prairie.

Time can change a prairie, though. We all know about Time. The first thing I remember learning in geology class is that the earth is only 4 billion years old, and at that they had to reach, and dig hard, all the way over in Australia, in one small location, to set that record. Oh, I think they even nudged it up to 4.3 bill. Which surprised me: I would have guessed about 10, or 95. If Methuselah with his own frail system of human pipes and tubing could move around on the earth for 969 years, almost a thousand, surely the earth, with Tetons (mountains up to 13,747 feet that stand storms of 100 mph) and Himalayas (average elevation of over 20,000 feet—do you know what an awful lot of rock that is? Try to move it, Methuselah) and ocean depths and lightning scratches and volcanoes . . . I juggle the power and immensity of not the earth but rather her forces, and take the square root of five and blow on a blade of grass and come up with an estimate that the earth is at *least* 10 billion years old.

What amazes me about the earth's discovered age is not that it is so old but that it is so young.

And for those who would frown and think this statement somehow demeans the work of a God, their God, again I would be surprised. By the assumption that with Their Grand Entrance the creation is, quite naturally, over. Who knows where it is going? Why should today be the stopping point, the reference point for history? Are we at the end, the beginning, or the middle?

So, theologically, I have no truck or qualms calling the earth 60 zillion years old. The fact that it is here is the glory; not how long or short a time it took to create it. Not assuming that speed is glory; not at all.

Why do things happen to you when they do? The wind and rain and frost prefer to work on jagged, protruding edges, a rectangular boulder, rather than an already smoothed river stone lying flat in the grass.

It is good to stand up and feel the wind blowing against you.

My shorts are on. I'm seated at the desk on my back porch, and I'm tingling after a cold bath. There are about ten minutes of cool daylight left (July 8). The cicadas are calling; the swifts shift in their nest and occasionally chitter. The sunset is swirling and red; it makes me wish I was sailing, and I don't even know how to sail.

There are so many things I do not know how to do!

I can write. I can find oil. I can cut yards.

People are so vulnerable. Engineers have a term for the condition of the hole: it is called skin damage, and it measures, in some God-awful micronesian units, the amount of abuse of any particular spot in the well bore that is due to drilling; due to the flow of circulating mud up and down the hole; due to the flow of fluids out of the formation (gas, oil, and/or water under pressure); or due to damage from the formation's shifting (flowing sand, sloughing shale). In short, skin damage summarizes all the rough knocks and setbacks a well encounters. It's a sensitivity barometer. Finding oil is a good talent, but companies and engineers and the rest of that gang know about skin damage, so that when they complete a well (run pipe into the hole, cement it in place, shoot holes through [perforate] the pipe and into the formation—out flows the pay and heroism and energy, independence and victory) they will be able to compare the well's performance (what it is "making") with its potential (what it *should* be making, what it is capable of doing). If the skin damage is too high, they could be tapped into a large reservoir and not know it! They might never get that oil out if they've damaged the well bore too much!

If you've got a poor well but a good engineer, he might be able to look at logs and pressures and production and such, and calculate the skin damage and then tell you to try again, drill another well in that reservoir, and take better care of the hole this time. You don't ever want to write anything off in the oil business, particularly a producing well that used to be good, no matter how marginal it is at present, without fully evaluating it. Taking into account skin damage and about a thousand other things.

There are things you may not know about your well. You've got to be careful not to judge it too quickly or too harshly.

The air is cool on the back of my neck. I am completely dry now, but my hair is still wet and it smells good. There was sun in Montana. My nose is peeling. These are abstract reflections, but there are more important things in life than a Chosen Profession. I would like to take the next step and tug on my beard and tell you what one of those things is, but I am only twenty-eight and I do not quite know, yet. Not well enough to tell you. Perhaps it is the love of another, or others. Perhaps it is doing what you do best. I cannot tell you if it is science or art. It may be self. It may very well be self, and honesty, and nothing beyond, and then happiness follows from that.

We pulled carrots out of the ground on our friend Vern's

ranch in Wyoming. They came out of the soil easily and we rinsed them off with his garden hose and ate them as we drove away from his house. We didn't see any grizzly bears or even black bears, but we met new people and it was different, fresh, and new.

July 27. There are more baby birds. It's only been a few weeks. It's a second hatch.

I am drinking a Coke, my second today, just fixed a grilled cheese too, to watch the birds with. No real need to hoard the Cokes anymore; Coke Classic is on the way, a return.

Last Wednesday, July 24, the first haze hit Jackson. The contour line that includes geographically similar towns and cities received roughly the same amount of sun, rain, wind, and lightning at the same time of the year. I think it's the cottonwoods along the rivers, creeks, and bayous that release this haze in the air.

It is a straight line from Jackson to Elizabeth's house on the Big Black. Nothing changes in those forty-three miles. The leaves, temperature, thickness in the air, scent, and thoughts: it is the same all the way there, along the same contour line.

—Still, the wary mind mistrusts these Big Companies. The Big People might have messed it up, this Coke repeat; it might never be recaptured. Elizabeth and I are awaiting the Coke Classics rather guardedly. We have heard all this talk about them but we haven't seen any yet, and we sure haven't tasted any.

When I someday drill it will be roughly fifteen miles north and east of Vernon, across the Lamar County line, into the high hills (six-hundred-foot elevations rather than the usual two hundred to three hundred feet) and red clay banks and

pine trees and stars at night that seem to discipline you, they are so severe and crisp. And I'll have on jeans and perhaps boots or tennis shoes and maybe a jacket, the night the drillers finish and the well's ready to log, maybe in the fall, and I'll step up on the logging truck and crouch down holding a cup of coffee, and the logging engineer will be seated at the computer in the back of the van-truck, and the little green screen will wink and blip and flash, and lines will begin racing up the screen, as if consuming it, like things breaking for cover, like a fire, filling in the blankness of the green screen with knowledge, yellow electronic lines of intelligence, and the Carter sandstone will be captured, identified, quantified, locked into. (Two hundred and fifty million years old.)

And it'll be oil. Because I agreed with my heart, didn't just take a chance but truly saw on the maps and with my heart the outlines and borders of the sand and the oil and believed what was within. The engineer will be glad too, because it means the existence of a new field, development drilling, more wells and ever more jobs, more security. He'll be friendly; mean people don't last in the logging business. The hours are too long, the job too hard, the anger or mean streak too meaningless and made insignificant by the exhaustion.

So everyone's happy. But I digress. (Back porch. Saturday, breeze blowing hard, dogs asleep.)

There can be oil indicated on the log, oil proven, certified, according to your past experience with the logs. But it is like the Coke Classic. There is always something that can go wrong. There are simple and tested procedural steps to go through, but you can't get cocky with the earth and *assume* there is nothing that will go wrong. When the oil is coming out and being sold and I am cashing the checks and going down to the Mayflower to buy redfish and coffee and pie with the money, even then I will not assume, but instead will be thankful to the two-hundred-and-fifty-million-year-old rock half a mile down there that I did find oil.

You have found the oil: now you just have to wait and see if it will come out, if the engineers can get it out. The dependence on them tastes like a fish hook.

Pacing, pacing. It is all a matter of timing, and knowing when to leap. I have seen geologists howl, ignited by greed, passion, lust, and foolishness, not to mention disrespect of the earth and the laws of nature, whims of happenstance. I've seen them run through the halls shouting, "Oil! Oil! Look at me!"

The look on their faces, after leaping too soon, is very akin to the look of someone who's stepped in something. Eating crow is much too mild a description of what they must feel. Two hundred and fifty million years is such a long time. Surely someone could keep quiet just one more day, until he or she is sure.

In my younger years, I was going to find a bumper sticker and stealthily paste it on the back of Mr. H.'s limousine, something in orange and blue like "Follow Me to Daiquiri World" or "I'm a Lover, a Fighter, and a Wild Bull Rider." Where does the energy of youth—constructive, destructive restlessness—go? Does it vaporize as age comes on? Where is it trapped? Against what caverns, under what sort of cap rock does it finally crawl off to and hide? What would happen if you found it? Could you get it out, produce it, or only look at it?

Full moon tonight. Sixty-five degrees. I am in Mississippi. Names of counties in my state: Tippah. Tishomingo. Tunica, Itawamba. Choctaw. Oktibbeha.

Mr. H. is sixty years old, famous, revered, and a billionaire. Sometimes, in my mind, I will accuse him of going for the headlines. I can understand this reaching for things more permanent and meaningful than the long-surpassed money goal. He is stalking more respect now. This part of being a man or a woman is another book, but I am twenty-eight, have a bad neck, and walk around the house in cut-offs. I need to drill all the oil, every bit of it; not just the slam-bang big remaining headliner reservoirs. The man could, given his money, assassinate the Black Warrior Basin. It confuses me that he does not, but I do not judge him for it.

At any rate, I made $5,000 of my salary last Friday by realizing we had gotten a permit to drill a well at Bluff down to Lewis depth, though since the permit was issued another company had drilled a Lewis offset that had pay. After running pipe on it, however, this other company soon lost pressure—it was pay, all right, but low volume, uneconomical (on the log you can't tell this). So, when I called everyone up late Friday night to get permission to drill only to Bangor lime, I might have saved $150,000, which is what it would have cost if we'd seen that same pay in the Lewis and gotten faked out and jacked around with it ourselves.

This is the sound of Rick blowing his own horn. But I have to, this well-sitting weekend. No one will ever know how much money I saved the company. Sometimes that stings. Some fields sting more than others; this Bluff field is one of them. I found the oil, delineated its thickness and outlined its presence—we're playing off some odd little facts on the density-neutron log and a fault I discovered in a then-obscure wildcat high up in the Pennsylvania section five years ago—and for two years, they wouldn't even let me change our regional map to show it, they weren't sure it existed, didn't have proof it carried for any great distance. Neither did I, but you have to try. You can always say no. And now it is the second largest oil field in the basin. I am still wasted from the last death-run (the Trace will kill you in a minute after ten P.M. Beware the Trace) and will skip a trip to see my grandparents in Fort Worth this weekend because a little while after it gets dark tonight I am going to have to crank up and head back to the basin again. I'll look at the samples all night, and then tell them when to shut down—stop drilling—and when to come out of the hole with the drill pipe and go in with the logging tools. I'll nap with my pinched neck nerves in the cab of the truck, under the stars, the pines. God, I love that country.

And when the sun gets hot we'll have the first log out. And then as the sun's going down I'll be starting back for Jackson.

Done repeatedly, it'll knock the pee out of you. Sooner or later something will go wrong.

It's the closest you can get to the oil. Sometimes when you find it, or are even the first to see it, it doesn't seem natural to come away without it.

———

My shoulder hurts. I will drill wells myself. This is non-sense. The mystery of it is that I still love it. It does me good to see people like Walter Sistrunk, who's been drilling the basin since who knows, the forties, who was the *only* person to drill it for about thirty years, and who still has the enthusiasm of a child.

I come around a dirt road too fast and swerve to miss the logging truck that is also coming around the corner, also too fast, about forty tons heavier, me laughing in the slide and waving, them laughing and waving too, also still sliding, good to see each other again. It's an exciting reunion; you depend on the loggers and they depend on you. Talking to all the other people who truly *do* own the well for those critical five or six hours when the little wells are being logged and the decisions are flying left and right. During that time, you really *do* care as much as they do, and you forget, every time, that none of the oil is yours; not any, even if you found it.

When you first see the oil and are running logs and deciding what to do to catch it, the ownership doesn't matter. It's a pure feeling.

It's when you're driving home, tired and empty-handed, that sorrow runs up and scratches at your door. You stop for gas at four A.M. and there's nothing else out and your windshield is splattered with dust and bugs and you need a quart of oil or have a flat tire or even two (bad roads).

You have to fight to stay young. Everybody, no exceptions, has to do that.

154

People who like material things are just fine by me—I mean, if they really like something, like the whole and all of the thing and not only the end result or money from it.

There are people in my industry who like the picture they make doing something, rather than the thing itself. Or else who made D's and F's in college and then married the boss's daughter and became insufferable. They learned to lie, or to aim other people's work in their direction. Then there are people who cannot do their job, cannot quite understand it, and are bitter and resentful because of this, and spend their hours plotting, whining, doing everything but the thing they should be focusing on, which is finding oil. Strong words, but oh, some days the oil business is a den of snakes, reptiles, and ghouls. On those days, the best thing to do is just to duck your head and drill the oil, get the oil. Eat peaches, strawberry waffles for breakfast. Kiss your girlfriend, kiss your lover. Get the oil. (This goes for any profession.)

When your writing is not going wonderfully, everything is suspect. It has to survive everything. Anything that's dear to me now will someday fall away because of my writing.

There's a Bug-Away machine on my kitchen floor. It has a long cord that trails across the floor and up over the counter and into the socket. It sits in the middle of the room, and if you bend down and listen very closely you can hear it ticking quietly. It was a gift from Elizabeth's mother, Janie, and the theory is that it emits high-pitched sonic waves that will rout fleas, mice, roaches, and silverfish but will not bother your pets. I do not have the nerve to tell you how much the machine cost.

I will sit in the other room and try to write, or try to map: to imagine where great unbounded stretches of oil are hiding. Or think up a story about a kid with his grandfather on a trip. A girl who works at the concession stand at an alligator farm. But nothing comes. For a fact, there are no mice, and Homer and Ann are in bliss, no scratching, no more fleas, but my writing seems stifled. The walls are ricocheting and echoing with these sonic waves that I cannot hear. I unplug the machine, open the windows; it is warm and windy outside, early evening. I love my dogs. It would be pleasurable not to have to trap mice, too; if they would just leave. Still, there is my writing. It comes first.

I have done so many more worse things than giving up the flea battle that it seems like only a yawn or a stretch compared with the discomfort I have caused others at times, just because I wanted to go down inside the strange place you have to go to be inside a story, on a seashore so many millions of years ago by yourself. The dogs trot out to the mailbox the next day, tails wagging, not knowing they have been betrayed.

And yet, things are back on the constant level. The dogs know I am going to open the lid of the mailbox and pull something out. They know that if I leave I will come back.

There are two ways to write: the way I do, and the way I want to. Sometimes people like what I write. The smartest readers know that I am saying nothing, but like a wild fighter, occasional punches slip through the defense: knockdown. I skate around the edges of mystic things: childhood, friendship, ponies, love—sketch them; no, *detail* their every line, so that friends and others say, Look! There it is! That's more development geology, defining boundaries as perfectly as possible, but then doing something about it—*entering* the reservoirs. That's where the childhood part must stop and the blood-and-guts part of the real world begins. It is the same with happiness. It is very easy to see the things that make you happy—define, sketch them—but to go *into* them, just duck your head and do it, quit your job (marry this girl instead of that one, not marry at all, work overtime, do a thousand sit-ups a night)—that's the difference between writing one way and writing another.

My friend John, a high school teacher in a little town in upstate New York, is the best teacher I have ever known. He's a strength coach, played football (center for Utah State while I was there), has a hilarious, clever brother, Jerry, and about a dozen beautiful sisters. He's not only a great teacher of athletics (I had never even perspired until I met him), but of everything he knows: carburetors, diapers, pancakes. I think it's because he goes to the interior. He doesn't just define; instead, he throws his hands up, gives a rough approximation of size and shape, and then enters. The good ones, the true students, follow.

Inside, he defines and sketches and details laboriously, expertly. Passionately. That's what I must do, with drilling, writing, Elizabeth. That's what everyone has to do, to even *approach* being happy with their lives.

John is one of the happiest people I know.

Later. Hurricane Elena landing in the state. We are expecting 90 mph winds. The horses are hiding in the barn. The dogs are, needless to say, frisky. Little hurricane dogs. The big barrier islands—Horn, Cat, Ship, Petit Bois, Dauphin—will be considerably rearranged; after today, the maps will have to be changed. A lot of times in the fossil record you will encounter, in a marine environment, what they call "rip-up clasts," chunks and wads of odd things— tree limbs, big rocks, and, in eons future, perhaps, beer cans, tractors—that would not normally be in that environment but that were displaced as the result of a single Big Event: a tornado, a hurricane, a thousand-year or ten-thousand-year flood.

Sometimes when I find myself inexplicably on a dance floor I feel a whole lot like a rip-up clast. How did I get out here? What am I doing here? Sometimes I meet great people, people I had no idea I would ever encounter anything like, and then, too, I feel like a rip-up clast, a grinning, fortunate rip-up clast. Fortune fuddles me. The only explanation I have ever heard that comes close to defining it is: Luck is the residual of preparation. In geology it is called serendipity. I think it's best not to rely on it. When it comes your way, hold on to it tighter than you've ever held anything. Do not let it go until it is gone.

Oil's found not only in sandstone, not only in sedimentary rocks. It's in cavernous limestones, old coral reefs, in metamorphic rocks such as shale—rocks under such pressure and twisted so intensely that their very identity was altered. Out in Nevada, you can even find oil and gas in some rule-defying porous igneous rocks. I know more about sands, as is probably true of the majority of petroleum geologists in this country. But you can find oil anywhere, if there was once life there, or at least nearby.

Geologists call long-ago happenings "events." This is an understatement. "Events" like the Gulf of Mexico, long ago, deciding one day (or one millennium) to come just a little farther ashore (warming trend on earth, melting ice caps), to, say, downtown Selma, Alabama. A transgression, we call it. And then, the regression: the Gulf can't hold its position. The earth's climate changes again, because of volcanic activity as North America drifts across a hot spot like the one present-day Hawaii covers, because of cosmic events, meteor showers, atmospheric alterations. Then the water pulls back. It leaves all sorts of ramps and flats and ocean-side deserts and dunes and swamps and evaporites and sinks and rises—it's rarely a simple draw-back, but instead a series of minor advances and retreats whose *overall* effect is backward, back into the big ocean. The water leaves sand and lime behind as it goes. Quite naturally, the basin areas—areas surrounded by high points such as mountain ranges and hills—are where the seas have advanced into and where the sand and lime are deposited. Sand is washed down from the hills and mountains and out of other basins, through rivers and streams and high winds, and into the oceans, gulfs, and bays, where it is shaped and positioned: rolled around and slung back up on shore or pulled farther along with the regression, pulled along with the ocean until the ocean is too weak to hold the weight (the sand's energy is always greatest closest to shore, where the waves hit).

Deltas, coming down out of the highlands, will chase, or be chased by, the movement of the ocean. If the ocean's backing away, over the years, the delta will follow, moving farther out into it, the finest, siltiest sands and clays chasing the retreating ocean and riding farthest out on the delta, farthest into the sea; not requiring as much energy to keep them in suspension as do the larger sand grains. (Mere memories of mountains; what was once footing for elk will someday be beneath starfish.)

And the finer grains pack tighter, closer to each other, becoming less porous, and are not as capable of holding much oil and gas. Oftentimes on a log you can see this deltaic response, just from an eight-inch hole (a mile or more deep, true) in the ground. The log will show that the sand is clean, at first—good and porous, with no silt at all—but then it will be less porous, siltier. (Ocean moving in? Is this well on the far reaches, the silty reaches of the delta now?) Then the sand may gradually increase in porosity again. (Ocean moving away from the delta; the delta chasing it, extending.)

Beyond calculating from the location of the deltas, of course, there is physics, and rough predictability: tides, pull of gravity, longshore currents, offshore and nearshore bars, barrier islands. A ramp heading downward, then flattening, then more ramping, then a drop-off—a flex, where the water goes rather suddenly from shallow to deep. It's more expensive to drill wells here, you have to go farther down, but often there is increased faulting along this flexure, and oil and gas will be trapped wherever it is porous enough (sand, in my basins) for them to hide in, all up and down these fault systems. Ramping appears to be behind the success of the more recent fields being discovered in the deeper part of the Black Warrior Basin: southern Lamar and Pickens counties. Beyond these variables there is scientific probability and confidence.

A new basin? An uplift, reefs? Minibasins and knolls? Development slows as the wells creep southward, deeper and deeper. We keep wanting (rightly so) to go back updip, back to the familiar, the proven, the less costly; less risks.

But also, slowly, wells venture into the new areas. Every success will breed three or four more tentative surrounding wells. A success in one of those three or four will breed maybe four or five more . . . Then no success, perhaps, and a mad dash, back we go into the shallows, the northern part of the county, burned and mistrustful, or maybe, like schooling minnows, the focus shifts to a new perimeter, a new theory, a new frontier—Marion County? The east flank of the basin, or maybe way north and into a new, unknown back basin, a lagoon sort of thing . . . It's all underground, and it's all very expensive, and you are reconstructing history from hundreds of millions of years ago with eight-inch circles in the ground, essentially sending blind men into the ancient, lost country and trying to fill in hundreds and hundreds of miles of buried forests and rivers and seas and dunes with these tiny pin sticks, like flag pins on a golf course. No, better, like trying to map the state of Colorado based only on what you can see for eight inches from, say, any fireplug in the state.

But you do not think of it this way. It will shake your confidence, your hope. Study the present-day Mississippi River, study today's Gulf of Mexico. These bodies of water don't change too much. Certainly gravity, physics, and geology (the rules, though not the results) don't alter much.

Don't think of that fireplug analogy at all.

You can always find a reason not to do something, or to be skeptical, or frightened.

There is no talent involved in not doing. You have to try to avoid dry holes; but to be frightened of them, frightened into inactivity and negativeness, means you have been defeated.

They're not alike at all, really: writing and geology. There's a deceit in writing; you're trying to pull all the clever elements together and toss out the dull and round-edged ones. Basically, it's building a lie and then swinging the lie's massiveness into the path of the reader and hiding behind it. Curiously, however, in geology, when I pour a cup of coffee and sit down and begin to map, I'm not hiding behind anything; there's no pretense, no deceit, just an inquisitive hunger and innocence where I am neither superior nor inferior to the reader, but *am* the reader. There's truly an amount of trust. The earth lies there, still, and obeys certain rules. I have faith that I am not going to let myself believe something that is not true. It is perhaps the purest thing I've ever done. Perhaps that is why geologists become so fervent about a particular prospect. Not holy men, but still there is that aspect to it—as in athletics, and religions.

Driving the long way around to Elizabeth's, green and purple summer dusk, down some new back roads. On the way to make up. God, how the heart quickens. It is like adolescence: it is that strong. Passing Tom Cain Road now; a bend, a hill, down through a creek bottom, and a shock of red cleared clay in the pines appears, a location pad for a drilling rig. In Warren County? Always it's a surprise to come upon one; like a fellow traveler in a foreign land, you both know the same language. You can't ever get away from the oil business. You can't leave it behind on Saturdays and Sundays, holidays. You're not in *it*, the business—it's a thing that's in *you*. If you like discovering and finding things, it's in you.

This farm is meant for me. I was watching a pine cone at the end of a branch this morning, and with no wind it fell. Just dropped, and bounced. I did not go over to pick it up; I was frightened to touch it or look at it. There was no telling what it could mean. I went and put on my tennis shoes and ran through the field and down through the woods until I was sweaty and breathing hard. I came out on a gravel road, and sat down, and panted. A white dog from someone's farm came trotting up. I petted him. Then I walked out to the main road, came back around to my house the long way, and things seemed better. I took a bath in the big old claw-footed tub I have and shaved and combed my hair and drove into Terry and bought a can of Gatorade and two bananas.

Still, I did not stare too long at anything on a tree, anywhere, for quite some time; I averted my glances, as if it were a question of modesty.

———————

I am fascinated with leaves these days. It is fall, and they are turning nicely this year. They give the trees, plum and pear and apple and oak, hickory, cottonwood, elm and willow, what seems to be the most tempting of messages. I want to stare, I feel like there's something I'm missing that's right in front of my face, but also if I stare too long and still fail to see it, the leaves will give up on me and stop trying to show me. So I watch out of the corners of my eyes. I feel very humbled.

Winter will come. I'll build fires over at Elizabeth's mother's house. I'll forget the leaves for another year. We will make spiced tea and popcorn. Trees should be trees, and people, people. I don't know why some people get sad in the fall, even while claiming paradoxically that it is their favorite time of year.

I should have put the pine cone in a mayonnaise jar and screwed the lid on and kept it next to my bed, or on the dashboard of my truck, for magic.

I was frightened.

There is a territoriality, a distance, that sets us apart from each other, and I do not know if that is good or bad. I don't know whether to hold on or let go. I don't think any of us ever knows, right up until the last moment.

This is from my freshman biology text, Utah State University:

> We vowed less than a decade ago that we would put a
> man on the moon and bring him back by the end of the
> 1960's. We met that deadline. Now maybe we should
> make a pledge that by the end of the 1970's we'll do some-
> thing really spectacular—like putting a man into Lake
> Erie and bringing him back out alive.
>
> —Senator J. Caleb Boggs

The book begins with the architecture of the cell, and 1,005 pages later concludes with the social behavior of humans. The book's been flooded by the Pearl River and also run over by my car but I still have it. I have to carry it with two hands. Its smell is distinctive. There is a word I don't know on every page. I thumb through it about once every month. To do so more often would make me sad, I think, for there is so much I would love to know intimately, as intimately as I know the Black Warrior Basin, marine bars, Elizabeth, favorite authors.

There just isn't going to be enough time. I don't want to be reminded of it.

The breeze, as ever, almost every night. And the crickets. I have to confess that I take them for granted. There is a little slip of a moon in the dark sky out the big open bedroom window; it is barely above eye level. It is sinful and wasteful to take things for granted, but guilt can be overdone, and I do not harbor any harshness for the people of the thirties, forties, fifties, sixties, and seventies who used Too Much Gasoline, Too Much Energy. (They were no different from the people of the eighties, of course. There is just a higher price and a little more awareness, a little less— though I do not accept it—oil.) There is a good taking for granted and a bad taking for granted, and at least theirs was good then: fun, travel, pleasure, quick energy.

The wanton polluters, the industries headed by people not in touch with their earth, my earth, ours . . . That's the bad taking for granted, the harming kind, and worse yet, the harming-others-for-one's-own-good kind.

The moon is rising high now. My cotton sheets are crisp. My lamp's making a yellow glow in the room. It is 9:46 and I can read awhile if I feel like it. The dogs are asleep on the back porch, curled up on top of each other, exhausted.

I guess I don't take it for granted after all.

Coffee. Cold sun. Dogs still in their box with their blanket, it's so cold; air down from Colorado, and me padding around in down booties in the quiet house. Crows calling like mad. Sky as blue as if it were Sunday. I think what I have discovered is that a state of mind, if you mold it right, is as real and durable as anything else capable of being retained in this world. Rock, earth, tree, mountain, seashore: good things last longer than bad. The farm is making me happy this morning without my even trying. It is lifting me up and carrying me. I created it, and it exists.

This is what I was looking for.